MEMOIRS
of the
American Mathematical Society

Number 1053

Fixed Point Theorems for Plane Continua with Applications

Alexander M. Blokh
Robbert J. Fokkink
John C. Mayer
Lex G. Oversteegen
E. D. Tymchatyn

July 2013 • Volume 224 • Number 1053 (second of 4 numbers) • ISSN 0065-9266

American Mathematical Society
Providence, Rhode Island

Library of Congress Cataloging-in-Publication Data

Blokh, Alexander M., 1958–
  Fixed point theorems for plane continua with applications / Alexander M. Blokh, Robbert J. Fokkink, John C. Mayer, Lex G. Oversteegen, E. D. Tymchatyn.
       p. cm. — (Memoirs of the American Mathematical Society, ISSN 0065-9266 ; no. 1053)
  "July 2013, volume 224, number 1053 (second of 4 numbers)."
  Includes bibliographical references and index.
  ISBN 978-0-8218-8488-1 (alk. paper)
  1. Fixed point theory. I. Title.
QA329.9.B56   2013
515'.39–dc23   2013006837

Memoirs of the American Mathematical Society

This journal is devoted entirely to research in pure and applied mathematics.

Publisher Item Identifier. The Publisher Item Identifier (PII) appears as a footnote on the Abstract page of each article. This alphanumeric string of characters uniquely identifies each article and can be used for future cataloguing, searching, and electronic retrieval.

Subscription information. Beginning with the January 2010 issue, *Memoirs* is accessible from www.ams.org/journals. The 2013 subscription begins with volume 221 and consists of six mailings, each containing one or more numbers. Subscription prices are as follows: for paper delivery, US$795 list, US$636 institutional member; for electronic delivery, US$700 list, US$560 institutional member. Upon request, subscribers to paper delivery of this journal are also entitled to receive electronic delivery. If ordering the paper version, add US$10 for delivery within the United States; US$69 for outside the United States. Subscription renewals are subject to late fees. See www.ams.org/help-faq for more journal subscription information. Each number may be ordered separately; *please specify number* when ordering an individual number.

Back number information. For back issues see www.ams.org/bookstore.

Subscriptions and orders should be addressed to the American Mathematical Society, P. O. Box 845904, Boston, MA 02284-5904 USA. *All orders must be accompanied by payment.* Other correspondence should be addressed to 201 Charles Street, Providence, RI 02904-2294 USA.

Copying and reprinting. Individual readers of this publication, and nonprofit libraries acting for them, are permitted to make fair use of the material, such as to copy a chapter for use in teaching or research. Permission is granted to quote brief passages from this publication in reviews, provided the customary acknowledgment of the source is given.

Republication, systematic copying, or multiple reproduction of any material in this publication is permitted only under license from the American Mathematical Society. Requests for such permission should be addressed to the Acquisitions Department, American Mathematical Society, 201 Charles Street, Providence, Rhode Island 02904-2294 USA. Requests can also be made by e-mail to reprint-permission@ams.org.

Memoirs of the American Mathematical Society (ISSN 0065-9266 (print); 1947-6221 (online)) is published bimonthly (each volume consisting usually of more than one number) by the American Mathematical Society at 201 Charles Street, Providence, RI 02904-2294 USA. Periodicals postage paid at Providence, RI. Postmaster: Send address changes to Memoirs, American Mathematical Society, 201 Charles Street, Providence, RI 02904-2294 USA.

© 2012 by the American Mathematical Society. All rights reserved.
Copyright of individual articles may revert to the public domain 28 years after publication. Contact the AMS for copyright status of individual articles.
This publication is indexed in *Mathematical Reviews*®, *Zentralblatt MATH, Science Citation Index*®, *Science Citation Index*^{TM}*-Expanded, ISI Alerting Services*^{SM}, *SciSearch*®, *Research Alert*®, *CompuMath Citation Index*®, *Current Contents*®/*Physical, Chemical & Earth Sciences*. This publication is archived in
Portico and *CLOCKSS*.
Printed in the United States of America.

∞ The paper used in this book is acid-free and falls within the guidelines
established to ensure permanence and durability.
Visit the AMS home page at http://www.ams.org/

10 9 8 7 6 5 4 3 2 1   18 17 16 15 14 13

Dedicated to Harold Bell

Contents

List of Figures	ix
Preface	xi
Chapter 1. Introduction	1

Part 1. Basic Theory — 11

Chapter 2. Preliminaries and outline of Part 1 — 13
  2.1. Index — 13
  2.2. Variation — 14
  2.3. Classes of maps — 15
  2.4. Partitioning domains — 17

Chapter 3. Tools — 19
  3.1. Stability of Index — 19
  3.2. Index and variation for finite partitions — 20
  3.3. Locating arcs of negative variation — 23
  3.4. Crosscuts and bumping arcs — 25
  3.5. Index and Variation for Carathéodory Loops — 27
  3.6. Prime Ends — 28
  3.7. Oriented maps — 30
  3.8. Induced maps of prime ends — 32

Chapter 4. Partitions of domains in the sphere — 35
  4.1. Kulkarni-Pinkall Partitions — 35
  4.2. Hyperbolic foliation of simply connected domains — 38
  4.3. Schoenflies Theorem — 40
  4.4. Prime ends — 41

Part 2. Applications of Basic Theory — 47

Chapter 5. Description of main results of Part 2 — 49
  5.1. Outchannels — 49
  5.2. Fixed points in invariant continua — 50
  5.3. Fixed points in non-invariant continua – the case of dendrites — 50
  5.4. Fixed points in non-invariant continua – the planar case — 51
  5.5. The polynomial case — 52

Chapter 6. Outchannels and their properties — 55
  6.1. Outchannels — 55

6.2.	Uniqueness of the Outchannel	59
Chapter 7.	Fixed points	63
7.1.	Fixed points in invariant continua	63
7.2.	Dendrites	64
7.3.	Non-invariant continua and positively oriented maps of the plane	69
7.4.	Maps with isolated fixed points	74
7.5.	Applications to complex dynamics	84
Bibliography		91
Index		95

Abstract

In this memoir we present proofs of basic results, including those developed so far by Harold Bell, for the plane fixed point problem: does every map of a non-separating plane continuum have a fixed point? Some of these results had been announced much earlier by Bell but without accessible proofs. We define the concept of the variation of a map on a simple closed curve and relate it to the index of the map on that curve: Index = Variation + 1. A prime end theory is developed through hyperbolic chords in maximal round balls contained in the complement of a non-separating plane continuum X. We define the concept of an *outchannel* for a fixed point free map which carries the boundary of X minimally into itself and prove that such a map has a *unique* outchannel, and that outchannel must have variation -1. Also Bell's Linchpin Theorem for a foliation of a simply connected domain, by closed convex subsets, is extended to arbitrary domains in the sphere.

We introduce the notion of an oriented map of the plane and show that the perfect oriented maps of the plane coincide with confluent (that is composition of monotone and open) perfect maps of the plane. A fixed point theorem for positively oriented, perfect maps of the plane is obtained. This generalizes results announced by Bell in 1982.

A continuous map of an interval $I \subset \mathbb{R}$ to $\mathbb{R}$ which sends the endpoints of I in opposite directions has a fixed point. We generalize this to maps on non-invariant continua in the plane under positively oriented maps of the plane (with

---

Received by the editor April 8, 2010, and, in revised form, December 20, 2011.

Article electronically published on November 16, 2012; S 0065-9266(2012)00671-X.

2010 *Mathematics Subject Classification*. Primary 37C25, 54H25; Secondary 37F10, 37F50, 37B45, 54C10.

Key words and phrases. Plane fixed point problem, crosscuts, variation, index, outchannel, dense channel, prime end, positively oriented map, plane continua, oriented maps, complex dynamics, Julia set.

The first named author was partially supported by grant NSF-DMS-0901038.

The fourth named author was supported in part by grant NSF-DMS-0906316.

The fifth named author was supported in part by NSERC 0GP0005616.

Affiliations at time of publication: Alexander M. Blokh, Department of Mathematics, University of Alabama at Birmingham, Birmingham, Alabama 35294-1170; email: ablokh@math.uab.edu; Robbert J. Fokkink, Delft Institute of Applied Mathematics, TU Delft, P.O. Box 5031, 2600 GA Delft, Netherlands, email: R.J.Fokkink@tudelft.nl; John C. Mayer, Department of Mathematics, University of Alabama at Birmingham, Birmingham, Alabama 35294-1170, email: mayer@math.uab.edu; Lex G. Oversteegen, Department of Mathematics, University of Alabama at Birmingham, Birmingham, Alabama 35294-1170, email: overstee@math.uab.edu; and E. D. Tymchatyn, Department of Mathematics and Statistics, University of Saskatchewan, Saskatoon, Saskatchewan, Canada S7N 0W0, email: tymchat@math.usask.ca.

©2012 American Mathematical Society

appropriate boundary conditions). Similar methods imply that in some cases non-invariant continua in the plane are degenerate. This has important applications in complex dynamics. E.g., a special case of our results shows that if X is a non-separating invariant subcontinuum of the Julia set of a polynomial P containing no fixed Cremer points and exhibiting no local rotation at all fixed points, then X must be a point. It follows that impressions of some external rays to polynomial Julia sets are degenerate.

List of Figures

3.1 Replacing $f : S \to \mathbb{C}$ by $f_1 : S \to \mathbb{C}$ with one less subarc of nonzero variation.   22

3.2 Bell's Lollipop.   24

3.3 $\mathrm{var}(f, A) = -1 + 1 - 1 = -1$.   27

4.1 Maximal balls have disjoint hulls.   36

6.1 The strip $\mathfrak{S}$ from Lemma 6.1.2   56

6.2 Uniqueness of the negative outchannel.   60

7.1 Replacing the links $[a_{n(1)-1}, a_{n(1)}]$, ..., $[a_{m(1)-1}, a_{m(1)}]$ by a single link $[a_{n(1)-1}, a_{m(1)}]$.   73

7.2 Illustration to the proof of Theorem 7.4.7.   78

7.3 Illustration to the proof of Lemma 7.4.9.   83

7.4 A general puzzle-piece   86

Preface

By a *continuum* we mean a compact and connected metric space and by a *non-separating* continuum X in the plane $\mathbb{C}$ we mean a continuum $X \subset \mathbb{C}$ such that $\mathbb{C} \setminus X$ is connected. Our work is motivated by the following long-standing problem [**Ste35**] in topology.

Plane Fixed Point Problem: *"Does a continuous function taking a non-separating plane continuum into itself always have a fixed point?"*

To give the reader perspective we would like to make a few brief historical remarks (see [**KW91, Bin69, Bin81**] for much more information).

Borsuk [**Bor35**] showed in 1932 that the answer to the above question is yes if X is also locally connected. Cartwright and Littlewood [**CL51**] showed in 1951 that a map of a non-separating plane continuum X to itself has a fixed point if the map can be extended to an *orientation-preserving* homeomorphism of the plane. It was 27 years before Harold Bell [**Bel78**] extended this result to the class of *all* homeomorphisms of the plane. Then Bell announced in 1982 (see also Akis [**Aki99**]) that the Cartwright-Littlewood Theorem can be extended to the class of all holomorphic maps of the plane. For other partial results in this direction see, e.g., [**Ham51, Hag71, Bel79, Min90, Hag96, Min99**].

In this memoir the Plane Fixed Point Problem is addressed. We develop and further generalize tools, first introduced by Bell, to elucidate the action of a fixed point free map (should one exist). We are indebted to Bell for sharing his insights with us. Some of the results in this memoir were first obtained by him. Unfortunately, many of the proofs were not accessible. Since there are now multiple papers which rely heavily upon these tools (e.g., [**OT07, BO09, BCLOS08**]) we believe that they deserve to be developed in a coherent fashion. We also hope that by making these tools available to the mathematical community, other applications of these results will be found. In fact, we include in Part 2 of this text new applications which illustrate their usefulness.

Part 1 contains the basic theory, the main ideas of which are due to Bell. We introduce Bell's notion of variation and prove his theorem that index equals variation increased by 1 (see Theorem 3.2.2). Bell's Linchpin Theorem 4.2.5 for simply connected domains is extended to arbitrary domains in the sphere and proved using an elegant argument due to Kulkarni and Pinkall [**KP94**]. Our version of this theorem (Theorem 4.1.5) is essential for the results later in the paper.

Building upon these ideas, we will introduce in Part 1 the class of oriented maps of the plane and show that it decomposes into two classes, one of which preserves and the other of which reverses local orientation. The extension from holomorphic to positively oriented maps is important since it allows for simple local perturbations of the map (see Lemma 7.5.1) and significantly simplifies further usage of the developed tools.

In Part 2 new applications of these results are considered. A Zorn's Lemma argument shows, that if one assumes a negative solution to the Plane Fixed Point Problem, then there is a subcontinuum X which is minimal invariant. It follows from Theorem 6.1.4 that for such a minimal continuum, $f(X) = X$. We recover Bell's result [**Bel67**] (see also Sieklucki [**Sie68**], and Iliadis [**Ili70**]) that the boundary of X is indecomposable with a dense channel (i.e., there exists a prime end $\mathcal{E}_t$ such that the principal set of the external ray R_t is all of ∂X).

As the first application we show in Chapter 6 that X has a *unique outchannel* (i.e., a channel in which points basically map farther and farther away from X) and this outchannel must have variation -1 (i.e., as the above mentioned points map farther and farther away from X, they are "flipped with respect to the center line of the channel").

The next application of the tools developed in Part 1 directly relates to the Plane Fixed Point Problem. We introduce the class of oriented maps of the plane (i.e., all perfect maps of the plane onto itself which are the compositions of monotone and branched covering maps of the plane). The class of oriented maps consists of two subclasses: positively oriented and negatively oriented maps. In Theorem 7.1.3 we show that the Cartwright-Littlewood Theorem can be extended to positively oriented maps of the plane.

These results are used in [**BO09**]. There we consider a branched covering map f of the plane. It follows from the above that if f has an invariant and fixed point free continuum Z, then f must be negatively oriented. We show in [**BO09**] that if, moreover, f is an oriented map of degree 2, then Z must contain a continuum X such that X is fully invariant (so that X contains the critical point and $f|_X$ is not one-to-one). Thus, X bears a strong resemblance to a connected filled in Julia set of a quadratic polynomial.

The rest of Part 2 is devoted to extending the existence of a fixed point in planar continua under positively oriented maps established in Theorem 7.1.3. We extend this result to non-invariant planar continua. First the result is generalized to dendrites; moreover, it is strengthened by showing that in certain cases the map must have infinitely many periodic cutpoints.

The above results on dendrites have applications in complex dynamics. For example, they are used in [**BCO08**] to give a criterion for the connected Julia set of a complex polynomial to have a non-degenerate locally connected model. That is, given a connected Julia set J of a complex polynomial P, it is shown in [**BCO08**] that there exists a locally connected topological Julia set J_{top} and a monotone map $m : J \to J_{top}$ such that for every monotone map $g : J \to X$ from J onto a locally connected continuum X, there exists a monotone map $f : J_{top} \to X$ such that $g = f \circ m$. Moreover, the map m has a dynamical meaning. It semi-conjugates the map $P|_J$ to a topological polynomial $P_{top} : J_{top} \to J_{top}$. In general, J_{top} can be a single point. In [**BCO08**] a necessary and sufficient condition for the non-degeneracy of J_{top} is obtained. These results extend Kiwi's fundamental result [**Kiw04**] on the semi-conjugacy of polynomials without Cremer or Siegel points to all polynomials with connected Julia set.

Finally the results on the existence of fixed points in invariant planar continua under positively oriented maps are extended to non-invariant planar continua. We introduce the notion of "scrambling of the boundary" of a plane continuum X under a positively oriented map and extend the fixed point results to non-invariant continua on which the map scrambles the boundary. These conclusions are strengthened by showing that, under additional assumptions, a non-degenerate continuum must either contain a fixed point in its interior, or must contain a fixed point near which the map "locally rotates". Hence, if neither of these is the case, then the continuum in question must be a point. This latter result is used to show that in certain cases impressions of external rays to connected Julia sets are degenerate.

These last named results have had other applications in complex dynamics. In [**BCLOS08**] these results were used to generalize the well-known Fatou-Shishikura inequality in the case of a polynomial P (in general, the Fatou-Shishikura inequality holds for rational functions, see [**Fat20, Shi87**]). For polynomials this inequality limits the number of attracting and irrationally neutral periodic cycles by the number of critical points of P. The improved count involves classes of (weakly recurrent) critical points and wandering subcontinua in the Julia set.

The results in Part 1 of this memoir were mostly obtained in the late 1990's. Most of the applications in Part 2, including the results on non-invariant plane continua and the applications in dynamics, have been obtained during 2006–2009. Finally the authors are indebted to a careful reading by the referee which resulted in numerous changes and improvements.

Alexander M. Blokh

Robbert J. Fokkink

John C. Mayer

Lex G. Oversteegen

E. D. Tymchatyn

CHAPTER 1

Introduction

1.0.1. Notation and the main problem. We denote the plane by $\mathbb{C}$, the Riemann sphere by $\mathbb{C}^\infty = \mathbb{C} \cup \{\infty\}$, the real line by $\mathbb{R}$ and the unit circle by $\mathbb{S}^1 = \mathbb{R}/\mathbb{Z}$. Let X be a plane compactum. Since $\mathbb{C}$ is locally connected and X is closed, complementary domains of X are open. By $T(X)$ we denote the *topological hull* of X consisting of X union all of its bounded complementary domains. Thus, $U^\infty = U^\infty(X) = \mathbb{C}^\infty \setminus T(X)$ is the unbounded complementary component of X containing infinity. Observe that if X is a continuum, then $U^\infty(X)$ is simply connected. The Plane Fixed Point Problem, attributed to [**Ste35**], is one of the central long-standing problems in plane topology. It serves as a motivation for our work and can be formulated as follows.

PROBLEM 1.0.1 (Plane Fixed Point Problem). Does a continuous function taking a non-separating plane continuum into itself always have a fixed point?

1.0.2. Historical remarks. To give the reader perspective we would like to make a few historical remarks concerning the Plane Fixed Point Problem (here we cover only major steps towards solving the problem).

In 1912 Brouwer [**Bro12**] proved that any orientation preserving homeomorphism of the plane, which keeps a bounded set invariant, must have a fixed point (though not necessarily in that set). This fundamental result has found many important applications. It was recognized early on that the location of a fixed point should be determined if the invariant set is a non-separating continuum (in that case a fixed point should be located in the invariant continuum) and many papers have been devoted to obtaining partial solutions to the Plane Fixed Point Problem.

Borsuk [**Bor35**] showed in 1932 that the answer is yes if X is also locally connected. Cartwright and Littlewood [**CL51**] showed in 1951 that a continuous map of a non-separating continuum X to itself has a fixed point in X if the map can be extended to an *orientation-preserving* homeomorphism of the plane. (See Brown [**Bro77**] for a very short proof of this theorem based on the above mentioned result by Brouwer). The proof by Cartwright-Littlewood Theorem made use of the *index of a map on a simple closed curve* and this idea has remained the basic approach in many partial solutions.

The most general result was obtained by Bell [**Bel67**] in the early 1960's. He showed that any counterexample must contain an invariant indecomposable subcontinuum. Hence the Plane Fixed Point Problem has a positive solution for hereditarily decomposable plane continua (i.e., for continua X which do not contain indecomposable subcontinua). Bell's result was also based on the notion of the index of a map, but he introduced new ideas to determine the index of a simple closed curve which runs tightly around a possible counterexample. Unfortunately, these ideas were not transparent and were never fully developed. Alternative proofs of Bell's result appeared soon after Bell's announcement (see [**Sie68, Ili70**]). Regrettably these results did not develop Bell's ideas.

In 1978 Bell [**Bel78**] used his earlier result to extend the result by Cartwright and Littlewood to the class of *all* homeomorphisms of the plane. Then Bell announced in 1982 (see also Akis [**Aki99**] where a wider class of differentiable functions was used) that the Cartwright-Littlewood Theorem can be extended to the class of all holomorphic maps of the plane. The existence of fixed points for orientation preserving homeomorphisms of the *entire plane* under various conditions was also considered in [**Bro84, Fat87, Fra92, Gui94**], and the existence of a point of period two for orientation reversing homeomorphisms in [**Bon04**].

As indicated above, positive results require an additional hypothesis either on the continuum X (as in Borsuk's result where the assumption is that X is locally connected) or on the map (as in Bell's case where the assumption is that f is a homeomorphism of the plane). Other positive results of the first type include results by Hamilton [**Ham51**] (X is chainable), Hagopian [**Hag71**] (X is arcwise connected) Minc [**Min90**] (X is the continuous image of the pseudo arc) and [**Hag96**] (X is simply connected). Positive results of the second type require the map to be either a homeomorphism [**CL51, Bel78**], holomorphic (as announced by Bell) or smooth with non-negative Jacobian and isolated singularities [**Aki99**].

David Bellamy [**Bel79**] produced an important related counterexample. He showed that there exists a tree-like continuum X, whose every proper subcontinuum is an arc and which admits a fixed point free homeomorphism. It is not known if examples of this type can be embedded in the plane. Minc [**Min99**] constructed a tree-like continuum which is the continuous image of the pseudo arc and admits a fixed point free map.

1.0.3. Major tools. In this subsection we describe the major tools developed in Part 1.

1.0.3.1. *Finding fixed points with index and variation.* It is easy to see that a map of a plane continuum to itself can be extended to a perfect map of the plane. We study the slightly more general question, "Is there a plane continuum Z and a perfect continuous function $f : \mathbb{C} \to \mathbb{C}$ taking Z into $T(Z)$ with no fixed points in $T(Z)$?" A Zorn's Lemma argument shows that if one assumes that the answer is "yes," then there is a subcontinuum $X \subset Z$, minimal with respect to these properties. It will follow from Theorem 6.1.4 that for such a minimal continuum, $f(X) = X = \partial T(X)$ (though it may not be the case that $f(T(X)) \subset T(X)$). Here $\partial T(X)$ denotes the boundary of $T(X)$.

Many fixed point results make use of the notion of the index $\mathrm{ind}(f, S)$, which counts the number of revolutions of the vector connecting z with $f(z)$ for $z \in S$ running along a simple closed curve S in the plane. As is well-known, if $f : \mathbb{C} \to \mathbb{C}$ is a map and $\mathrm{ind}(f, S) \ne 0$, then f must have a fixed point in $T(S)$ (for completeness we prove this in Theorem 3.1.4). In order to establish fixed points in invariant plane continua X, one often approximates X by a simple closed curve S such that $X \subset T(S)$. If $\mathrm{ind}(f, S) \ne 0$ and S is sufficiently tight around X, one can conclude that f must have a fixed point in $T(X)$. Hence the main work is in showing that $\mathrm{ind}(f, S) \ne 0$ for a suitable simple closed curve around X.

Bell's fundamental idea was to replace the count of the number of rotations of the vector $\overrightarrow{zf(z)}$ with respect to a fixed axis (say, the x-axis) by a count which involves the moving frame of external rays. Consider, for example, the unit circle

$\mathbb{S}^1$ and a fixed point free map $f : \mathbb{S}^1 \to \mathbb{C}$. For each $z = e^{2\pi i\theta} \in \mathbb{S}^1$ let R_z be the external ray $\{re^{2\pi i\theta} \mid r > 1\}$. Now count the number of times the point $f(z)$, $z \in \mathbb{S}^1$, crosses the external ray R_z, taking into account the direction of the crossing. Call this count the variation $\text{var}(f, \mathbb{S}^1)$. It is easy to see that in this case $\text{ind}(f, \mathbb{S}^1) = \text{var}(f, \mathbb{S}^1) + 1$.

Another useful idea is to consider a similar count not on the entire unit circle (or, in general, not on the entire simple closed curve S containing X in its topological hull $T(S)$), but on subarcs of $\mathbb{S}^1$, which map off themselves and whose endpoints map inside $T(\mathbb{S}^1)$. By doing so, one obtains Bell's notion of variation $\text{var}(f, A)$ on arcs (see Definition 2.2.2). If one can write $\mathbb{S}^1$ as a finite union A_i of arcs such that any two meet at most in a common endpoint and, for all i, $f(A_i) \cap A_i = \emptyset$ and both endpoints map in $T(\mathbb{S}^1)$, one can define $\text{var}(f, \mathbb{S}^1) = \sum \text{var}(f, A_i)$. Then, as above, one can use it to compute the index and prove that index equals variation increased by 1 (see Theorem 3.2.2).

The relation between index and variation immediately implies a few classic results, in particular that of Cartwright and Littlewood. To see this one only needs to show that, if h is an orientation preserving homeomorphism of the plane, X is an invariant plane continuum and $A_i \subset S$ are subarcs of a tight simple closed curve around X, with endpoints in X as above and such that $f(A_i) \cap A_i = \emptyset$, then $\text{var}(f, A_i) \geq 0$ (see Corollary 7.1.2). Then $\text{ind}(f, S) = \sum \text{var}(f, A_i) + 1 \geq 1$ and $T(X)$ contains a fixed point as desired. Observe that the connection between index and variation is essential for all the applications in Part 2.

1.0.3.2. *Other tools, such as foliations and oriented maps.* Let X be a non-separating plane continuum and let $f : \mathbb{C} \to \mathbb{C}$ be a map such that $f(X) \subset T(X)$ and f has no fixed points in $T(X)$. In general we need more control of the simple closed curve S around X (and of the action of the map on $S \setminus X$). Bell originally accomplished this by partitioning the complement of $T(X)$ in the Euclidean convex hull $\text{conv}_\mathcal{E}(X)$ of X by Euclidean convex sets. Suppose that B is a maximal round closed ball (or a half plane) such that $\text{int}(B) \cap T(X) = \emptyset$ and $|B \cap X| \geq 2$, and consider the set $\text{conv}_\mathcal{E}(B \cap X)$. For any two such balls B_1, B_2 either $K_{1,2} = \text{conv}_\mathcal{E}(B_1 \cap X) \cap \text{conv}_\mathcal{E}(B_2 \cap X)$ is empty, or $K_{1,2}$ is a single point in X, or this intersection is a common chord contained in both of their boundaries. Bell's Linchpin Theorem (see Theorem 4.2.5 and the remark following it) states that the collection $\text{conv}_\mathcal{E}(B \cap X)$ over all such maximal balls covers all of $\text{conv}_\mathcal{E}(X) \setminus T(X)$.

Hence the collection $\text{conv}_\mathcal{E}(B \cap X)$ over all such balls provides a partition of $\text{conv}_\mathcal{E}(X) \setminus X$ into Euclidean convex sets contained in maximal round balls. The collection of chords in the boundaries of the sets $\text{conv}_\mathcal{E}(B \cap X)$ for all such balls have the property that any two distinct chords meet in at most a common endpoint in X. In other words, this set of chords is a lamination in the sense of Thurston [**Thu09**] even though in Thurston's paper laminations appear in a very different, namely complex dynamical, context. This Linchpin Theorem can be used to extend the map $f|_{T(X)}$ over $\text{conv}_\mathcal{E}(X) \setminus T(X)$ (first linearly over all the chords in the lamination and then over all remaining components of the complement). We will illustrate the usefulness of Bell's partition by showing that the well-known Schoenflies Theorem follows immediately. It can also be used to obtain a particular simple closed curve S around X so that every component of $S \setminus X$ is a chord in the lamination.

In our version of Bell's Linchpin Theorem we consider arbitrary open and connected subsets of the sphere U and we use round balls in the spherical metric on the sphere. Moreover, the Euclidean geodesics are replaced by hyperbolic geodesics (in either the hyperbolic metric in each ball or, if U is simply connected, in the hyperbolic metric on U). This way we get a lamination of all of $U^\infty(X) = \mathbb{C}^\infty \setminus T(X)$ (and not just $\text{conv}_\mathcal{E}(X) \setminus T(X)$) and the resulting lamination is easier to apply in other settings. We give a proof of this theorem using an elegant argument due to Kulkarni and Pinkall [**KP94**] which also allows for the extension over arbitrary open and connected subsets of the sphere (see Theorem 4.1.5; this later theorem is used in Chapter 6). Bell's Linchpin Theorem follows as a corollary.

This new partition of a complementary domain of a continuum can also be used in other settings to extend a homeomorphism on the boundary of a planar domain over the entire domain. In [**OT07**] this is used to show that an isotopy of a planar continuum, starting at the identity, extends to an isotopy of the plane. This extends a well-known result regarding the extension of a holomorphic motion [**ST86, Slo91**]. In [**OV09**] this partition is used to give necessary and sufficient conditions to extend a homeomorphism, of an arbitrary planar continuum, over the plane.

The development of the necessary tools in Part 1 is completed by introducing the notion of (positively or negatively) oriented maps and studying their properties. Holomorphic maps are prototypes of positively oriented maps but in general positively oriented maps do not have to be differentiable, light, open or monotone. Locally at non-critical points, positively oriented maps behave like orientation-preserving homeomorphisms in the sense that they preserve local orientation. Compositions of open, perfect and of monotone, perfect surjections of the plane are *confluent* (i.e., such that components of the preimage of any continuum map onto the continuum) and naturally decompose into two classes, one of which preserves and the other of which reverses local orientation. We show that any confluent map of the plane is itself a composition of a monotone and a light-open map of the plane. It is shown that an oriented map of the plane induces a map from the circle of prime ends of a component of the pre-image of an acyclic plane continuum to the circle of prime ends of that continuum.

1.0.4. Main applications. Part 2 contains applications of the tools developed in Part 1. Directly or indirectly, these applications deal with the Plane Fixed Point Problem. We describe them below.

1.0.4.1. *Outchannel and hypothetical minimal continua without fixed points.* The first application is in Chapter 6 where we establish the existence of a unique outchannel. Let us consider this in more detail. If there exists a counterexample to the Plane Fixed Point Problem, then there exists a continuum X which is minimal with respect to $f(X) \subset T(X)$ and f has no fixed point in $T(X)$. Bell has shown [**Bel67**] (see also [**Sie68, Ili70**]) that such a continuum has at least one dense outchannel of negative variation. Since X is minimal with a dense outchannel, X is an indecomposable continuum and $f(X) = X$.

A dense outchannel is a prime end so that its principal set is all of X and if $\{C_i\}$ is a defining sequence of crosscuts, then f maps these crosscuts essentially "out of the channel" (i.e., closer to infinity) for i sufficiently large. The latter statement is accurately reflected by the fact that $\text{var}(f, C_i) \neq 0$. In case that the complement of X is invariant, this can be described by saying that the crosscut $f(C_i)$ separates C_i from infinity in $U^\infty(X)$ (and hence, in this case, crosscuts do really map out of the channel). As a new result, the main steps of the proof of which were outlined by

Bell, we show that there always exists *exactly one outchannel* and that its variation is -1, while all other prime ends must have variation 0. Using these results it is shown in [**BO09**] that if f is a negatively oriented branched covering map of degree 2 which has a non-separating invariant continuum Z without fixed points, then the minimal subcontinuum $X \subset Z$ has the following additional properties:

(1) X is indecomposable,
(2) the unique critical point c of f belongs to X, $f(X) = X$, and $f(\mathbb{C} \setminus X) = \mathbb{C} \setminus X$ (so that X is *fully invariant*),
(3) f induces a covering map $F : \mathbb{S}^1 \to \mathbb{S}^1$ from the circle of prime ends of X to itself of degree -2.
(4) F has three fixed points one of which corresponds to the unique dense outchannel whereas the remaining two fixed points correspond to dense inchannels (i.e., for a defining sequence of crosscuts $\{C_i\}$, C_i separates $f(C_i)$ from infinity in U^∞)

Moreover, as part of the argument, the map f is modified in $\mathbb{C} \setminus X$ so that the new map g keeps the tail of the external ray, which runs down the outchannel, invariant and maps the points on them closer to infinity.

1.0.4.2. *Fixed points in invariant continua for positively oriented maps.* Other applications of the tools developed in Part 1 are obtained in Chapter 7. These are also related to the Plane Fixed Point Problem. As we will see below, the corresponding results can be in turn further applied in complex dynamics, leading to some structural results in the field, such as constructing finest locally connected models for connected Julia sets or studying wandering continua inside Julia sets and an extension of the Fatou-Shishikura inequality so that it includes counting wandering branch-continua (see Section 7.5).

The first application in Chapter 7 is the most straightforward of them all: in Theorem 7.1.3 from Section 7.1 we prove that a positively oriented map f which takes a continuum X into the topological hull $T(X)$ of X must have a fixed point in $T(X)$. In other words, in Theorem 7.1.3 the Plane Fixed Point Problem is solved in the affirmative for positively oriented maps. As we will see, the extension from holomorphic maps to positively oriented maps is important since the latter class allows for easy local perturbations. This will allow us to deal with parabolic points in a Julia set (see Lemma 7.5.1, Theorem 7.5.2 and Corollary 7.5.4).

The idea of the proof is as follows. First we prove in Corollary 7.1.2 that if a crosscut C of X is mapped off itself by f then the variation on C is non-negative. This is done by completing the crosscut C to a very tight simple closed curve S around X and observing that in fact the variation in question can be computed by computing the winding number of f on S. Notice that versions of this idea are used later on when we prove the existence of fixed points in non-invariant continua satisfying certain additional conditions.

To prove Theorem 7.1.3, we first assume by way of contradiction that $T(X)$ contains no fixed points. In this case there are no fixed points in the closure $\overline{U}$ of a sufficiently small neighborhood U of $T(X)$. Using this, we construct a simple closed curve which goes around X inside such a neighborhood U and "touches" X at a sufficiently dense set of points so that arcs between consecutive points of $S \cap X$ are very small. Since there are no fixed points in $\overline{U}$, we can guarantee that the images of these arcs are disjoint from themselves. Hence by the above described Corollary 7.1.2 the variations of all these arcs are non-negative. By Theorem 3.2.2

this implies that the index of f on S is not equal to zero and hence, by Theorem 3.1.4 there must exist a fixed point inside $T(S)$, a contradiction.

1.0.4.3. *Fixed points in non-invariant continua: the case of dendrites.* Our generalizations of Theorem 7.1.3 are inspired by a simple observation. The most well-known particular case for which the Plane Fixed Point Problem is solved is that of a map of a closed interval $I = [a, b]$, $a < b$ into itself in which case there must exist a fixed point in I. However, in this case a more general result can easily be proven, of which the existence of a fixed point in an invariant interval is a consequence.

Namely, instead of considering a map $f : I \to I$ consider a map $f : I \to \mathbb{R}$ such that either (a) $f(a) \geq a$ and $f(b) \leq b$, or (b) $f(a) \leq a$ and $f(b) \geq b$. Then still there must exist a fixed point in I which is an easy corollary of the Intermediate Value Theorem applied to the function $f(x) - x$. Observe that in this case I need not be invariant under f. Observe also that without the assumptions on the endpoints, the conclusion on the existence of a fixed point inside I cannot be made because, e.g., a shift map on I does not have fixed points at all. The conditions (a) and (b) above can be thought of as boundary conditions imposing restrictions on where f maps the boundary points of I in $\mathbb{R}$.

Our main aim in the remaining part of Chapter 7 is to consider some other cases for which the Plane Fixed Point Problem can be solved in the affirmative (i.e., the existence of a fixed point in a continuum can be established) despite the fact that the continuum X in question is not invariant. We proceed with our studies in two directions. Considering X, we replace the invariantness of the continuum by boundary conditions in the spirit of the above "interval version" of the Plane Fixed Point Problem. We also show that there must exist a fixed point of "rotational type" in the continuum (and hence, if it is known that such a point does not exist, then the continuum in question is a point).

Since we now deal with continua significantly more complicated than an interval, inevitably the boundary conditions become rather intricate. Thus we postpone the precise technical statement of the results until Chapter 7 and use here a more descriptive approach. Observe that particular cases for which the Plane Fixed Point Problem is solved so far can be divided into two categories: either X has additional properties, or f has additional properties. In the first category the above considered "interval case" is the most well-known. A direct extension of it is the following well-known theorem (which follows from Borsuk's theorem [**Bor35**], see [**Nad92**] for a direct proof); recall that a *dendrite* is a locally connected continuum containing no simple closed curves.

THEOREM 1.0.2. *If $f : D \to D$ is a continuous map of a dendrite into itself then it has a fixed point.*

Here f is just a continuous map but the continuum D is very nice. In Section 7.2 Theorem 1.0.2 is generalized to the case when $f : D_1 \to D_2$ maps a dendrite D_1 into a dendrite $D_2 \supset D_1$ and certain conditions on the behavior of the points of the set $E = \overline{D_2 \setminus D_1} \cap D_1$ under the map f are fulfilled (observe that E may be infinite). This presents a "non-invariant" version of Plane Fixed Point Problem for dendrites and can be done in the spirit of the interval case described earlier. Moreover, with some additional conditions it has consequences related to the number of periodic points of f.

More precisely, we introduce the notion of *boundary scrambling* for dendrites in the situation above. It simply means that for each *non-fixed* point $e \in E$, $f(e)$

is contained in a component of $D_2 \setminus \{e\}$ which intersects D_1 (see Definition 5.3.1). Observe that if D_1 *is* invariant then f automatically scrambles the boundary. We prove the following theorem.

THEOREM 7.2.2. *Suppose that $f : D_1 \to D_2$ is a map between dendrites, where $D_1 \subset D_2$, which scrambles the boundary. Then f has a fixed point.*

Next in Section 7.2 we define *weakly repelling periodic points*. Basically, a point $a \in D_1$ is a *weakly repelling periodic point (for f^n)* if there exists $n \geq 1$ and a component B of $D_1 \setminus \{a\}$ such that $f^n(a) = a$ and arbitrarily close to a in B there exist cutpoints of D_1 fixed under f^n or points x separating a from $f^n(x)$. Note that a fixed point a of f can be a weakly repelling periodic point for f^n while it is not weakly repelling for f. We use this notion to prove Theorem 7.2.6 where we show that if D is a dendrite and $f : D \to D$ is continuous and all its periodic points are weakly repelling, then f has infinitely many periodic cutpoints. Then we rely upon Theorem 7.2.6 in Theorem 7.2.7 where it is shown that if $g : J \to J$ is a *topological polynomial* on its dendritic Julia set (e.g., if g is a complex polynomial with a dendritic Julia set) then it has infinitely many periodic cutpoints.

1.0.4.4. *Fixed points in non-invariant continua: the planar case.* In Sections 7.3 and 7.4 we draw a parallel with the interval case for planar maps and extend Theorem 7.1.3 to non-invariant continua under positively oriented maps such that certain "boundary" conditions are satisfied. Namely, suppose that $f : \mathbb{C} \to \mathbb{C}$ is a positively oriented map and $X \subset \mathbb{C}$ is a non-separating continuum. Since we are interested in fixed points of $f|_X$, it makes sense to assume that at least $f(X) \cap X \neq \emptyset$. Thus, we can think of $f(X)$ as a new continuum which "grows" from X at some places. We assume that the "pieces" of $f(X)$ which grow outside X are contained in disjoint non-separating continua Z_i so that $f(X) \setminus X \subset \cup_i Z_i$.

We also assume that places at which the growth takes places - i.e., sets $Z_i \cap X = K_i$ - are non-separating continua for all i. Finally, the main assumption here is the following restriction upon where the continua K_i map under f: we assume that for all i, $f(K_i) \cap [Z_i \setminus K_i] = \emptyset$. If this is all that is satisfied, then the map f is said to *scramble the boundary (of X)*. A stronger version of that is when for all i, either $f(K_i) \subset K_i$, or $f(K_i) \cap Z_i = \emptyset$; then we say that f *strongly scrambles the boundary (of X)* (see Definition 5.4.1). The continua K_i are called *exit continua (of X)*. The main result of Section 7.3 is:

THEOREM 7.3.3. *Suppose that f is positively oriented and strongly scrambles the boundary of X, then f has a fixed point in X.*

As an illustration, consider the case when $X \cup (\cup_i Z_i)$ is a dendrite and all sets K_i are singletons. Then it is easy to see that both scrambling and strong scrambling of the boundary in the sense of dendrites mean the same as in the sense of the planar definition. Of course, in the planar case we deal with a much more narrow class of maps, namely positively oriented maps, and with a much wider variety of continua, namely all non-separating planar continua. This fits into the "philosophy" of our approach: whenever we obtain a result for a wider class of continua, we have to consider a more specific class of maps.

For the family of positively oriented maps with isolated fixed points we specify this result as follows. We introduce the notion of the map f *repelling outside X* at a fixed point p (see Definition 7.4.5; basically, it means that there exists an invariant external ray of X which lands at p and along which the points are repelled away from p. Then in Theorem 7.4.7 we show that *if f is a positively oriented map with isolated fixed points and $X \subset \mathbb{C}$ is a non-separating continuum or a point such that f scrambles the boundary of X and for every fixed point a the winding number at a equals 1 and f repels at a, then X must be a point.*

1.0.4.5. *Fixed points in non-invariant continua for polynomials.* These theorems apply to polynomials P, allowing us to obtain a few corollaries dealing with the existence of periodic points in certain parts of the Julia set of a polynomial and degeneracy of certain impressions. To discuss this we assume knowledge of the standard definitions such as *Julia sets* J_P, *filled-in Julia sets* $K_P = T(J_P)$, *Fatou domains, parabolic periodic points* etc which are formally introduced in Section 5.5 and further discussed in Section 7.5 (see also [**Mil00**]). Recall that the set $U^\infty(J_P)$ (called in this context the *basin of attraction of infinity*) is partitioned by the *(conformal) external rays* R_α with arguments $\alpha \in \mathbb{S}^1$. If J_P is connected, all rays R_α are smooth and pairwise disjoint while if J_P is not connected limits of smooth external rays must be added. Still, given an external ray R_α of K, its principal set $\overline{R_\alpha} \setminus R_\alpha$ can be introduced as usual.

We then define a *general puzzle-piece* of a filled-in Julia set K_P as a continuum X which is cut from K_P by means of choosing a few *exit continua* $E_i \subset X$ each of which contains the principal sets of more than one external ray. We then assume that there exists a component C_X of the complement in $\mathbb{C}$ to the union of all such exit continua E_i and their external rays such that $X \subset (C_X \cap K_P) \cup (\bigcup E_i)$. The external rays accumulating inside an exit continuum E_i cut the plane into wedges one of which, denoted by W_i, contains points of X. The "degenerate" case when there are no exit continua is also included and simply means that X is an invariant subcontinuum of K_P.

The main assumption on the dynamics of a general puzzle piece X which we make is that $P(X) \cap C_X \subset X$ and for each exit continuum E_i we have $P(E_i) \subset W_i$. It is easy to see that this essentially means that P scrambles the boundary of X (where the role of the "boundary" is played by the union of exit continua).

The conclusion, obtained in Theorem 7.5.2, is based upon the above described results, in particular on Theorem 7.4.7. It states that for a general puzzle-piece either X contains an invariant parabolic Fatou domain, or X contains a fixed point which is neither repelling nor parabolic, or X contains a repelling or parabolic fixed point a at which the local rotation number is not 0. Let us now list the main dynamical applications of this result.

1.0.4.6. *Further dynamical applications.* There are a few ways Theorem 7.5.2 applies in complex (polynomial) dynamics. First, it is instrumental in studying *wandering cut-continua* for polynomials with connected Julia sets. A continuum/point $L \subset J_P$ is a *cut-continuum (of valence* val(L)) if the cardinality val(L) of the set of components of $J_P \setminus L$ is greater than 1. A collection of disjoint cut-continua (it might, in particular, consist of one continuum) is said to be *wandering* if their forward images form a family of pairwise disjoint sets. The main result of [**BCLOS08**] in the case of polynomials with connected Julia sets is the following generalization of the Fatou-Shishikura inequality.

THEOREM 1.0.3. *Let P be a polynomial with connected Julia set, let N be the sum of the number of distinct cycles of its bounded Fatou domains and the number of cycles of its Cremer points, and let $\Gamma \neq \emptyset$ be a wandering collection of* cut-continua Q_i with valences greater than 2 which contain no preimages of critical points of P. Then $\sum_\Gamma(\mathrm{val}(Q_i) - 2) + N \leq d - 2$.

In [**BCLOS08**] a partition of the plane into pieces by rays with rational arguments landing at periodic cutpoints of J_P and their preimages is used. Theorem 7.5.2 plays a significant role in the proof of the fact that wandering cut-continua do not enter the pieces containing Cremer or Siegel periodic points which is an important ingredient of the arguments in [**BCLOS08**] proving Theorem 1.0.3.

Another application of Theorem 7.5.2 can be found in [**BCO08**] where Kiwi's fundamental result [**Kiw04**] on the semiconjugacy of polynomials on their Julia sets without Cremer or Siegel points is extended to all polynomials with connected Julia sets; in both cases topological polynomials on their topological Julia sets serve as locally connected models. Denote the monotone semiconjugacy in question by φ. In showing in [**BCO08**] that if x is a (pre)periodic point and $\varphi(x)$ is not equal to a φ-image of a Cremer or Siegel point or its preimage then J_P is locally connected at x, Theorem 7.5.2 plays a crucial role.

Finally, our results concerning dendrites (such as Theorem 7.2.6 and Theorem 7.2.7) are used in [**BCO08**] where a criterion for the connected Julia set to have a non-degenerate locally connected model is obtained. We also rely on Theorem 7.2.6 and Theorem 7.2.7 to show in [**BCO08**] that if such model exists, and is a dendrite, then the polynomial must have infinitely many bi-accessible periodic points in its Julia set.

1.0.5. Concluding remarks and acknowledgments. All of the positive results on the existence of fixed points in this memoir are either for simple continua (i.e., those which do not contain indecomposable subcontinua) or for positively oriented maps of the plane. Hence the following special case of the Plane Fixed Point Problem is a major remaining open problem:

PROBLEM 1.0.4. Suppose that $f : \mathbb{C} \to \mathbb{C}$ is a negatively oriented branched covering map, $|f^{-1}(y)| \leq 2$ for all $y \in \mathbb{C}$ and Z is a non-separating plane continuum such that $f(Z) \subset Z$. Must f have a fixed point in Z?

Suppose that c is the unique critical point of f and that $X \subset Z$ is a minimal continuum such that $f(X) \subset T(X)$. Then, as was mentioned above, the answer is yes if there exists $y \in X \setminus \{f(c)\}$ such that $|f^{-1}(y) \cap X| < 2$. In particular the answer to Problem 1.0.4 is yes if $f|_Z$ is one-to-one.

Finally let us express, once again, our gratitude to Harold Bell for sharing his insights with us. His notion of variation of an arc, his index equals variation plus one theorem and his linchpin theorem of partitioning a complementary domain of a planar continuum into convex subsets are essential for the results we obtain here. Theorem 6.2.1 (Unique Outchannel) is a new result the main steps of which were outlined by Bell. Complete proofs of the following results by Bell: Theorems 3.2.2, 3.3.1, 4.2.5 and 6.2.1, appear in print for the first time. For the convenience of the reader we have included an index at the end of the paper.

Part 1

Basic Theory

CHAPTER 2

Preliminaries and outline of Part 1

In this chapter we give the formal definitions and describe the results of part 1 in more detail. By a *map* $f : X \to Y$ we will always mean a continuous function.

Let $p : \mathbb{R} \to \mathbb{S}^1$ denote the covering map $p(x) = e^{2\pi i x}$. Let $g : \mathbb{S}^1 \to \mathbb{S}^1$ be a map. By the *degree* of the map g, denoted by degree(g), we mean the number $\hat{g}(1) - \hat{g}(0)$, where $\hat{g} : \mathbb{R} \to \mathbb{R}$ is a lift of the map g to the universal covering space $\mathbb{R}$ of $\mathbb{S}^1$ (i.e., $p \circ \hat{g} = g \circ p$). It is well-known that degree(g) is independent of the choice of the lift.

2.1. Index

Let $g : \mathbb{S}^1 \to \mathbb{C}$ be a map and $f : g(\mathbb{S}^1) \to \mathbb{C}$ a fixed point free map. Define the map $v : \mathbb{S}^1 \to \mathbb{S}^1$ by
$$v(t) = \frac{f(g(t)) - g(t)}{|f(g(t)) - g(t)|}.$$

Then the map $v : \mathbb{S}^1 \to \mathbb{S}^1$ lifts to a map $\hat{v} : \mathbb{R} \to \mathbb{R}$. Define the *index of f with respect to g*, denoted ind(f, g) by
$$\text{ind}(f, g) = \hat{v}(1) - \hat{v}(0) = \text{degree}(v).$$

Note that ind(f,g) measures the net number of revolutions of the vector $f(g(t)) - g(t)$ as t travels through the unit circle one revolution in the positive direction.

REMARK 2.1.1. The following basic facts hold.

(a) If $g : \mathbb{S}^1 \to \mathbb{C}$ is a constant map with $g(\mathbb{S}^1) = c$ and $f(c) \neq c$, then ind$(f, g) = 0$.

(b) If f is a constant map and $f(\mathbb{C}) = w$ with $w \notin g(\mathbb{S}^1)$, then ind$(f, g) = $ win$(g, \mathbb{S}^1, w)$, the *winding number of g about w*. In particular, if $f : \mathbb{S}^1 \to T(\mathbb{S}^1) \setminus \mathbb{S}^1$ is a constant map, then ind$(f, id|_{\mathbb{S}^1}) = 1$, where $id|_{\mathbb{S}^1}$ is the identity map on $\mathbb{S}^1$.

Note also, that for a simple closed curve S' and a point $w \notin T(f(S'))$ we have win$(f, S', w) = 0$. Suppose $S \subset \mathbb{C}$ is a simple closed curve and $A \subset S$ is a subarc of S with endpoints a and b. Then we write $A = [a, b]$ if A is the arc obtained by traveling in the counter-clockwise direction from the point a to the point b along S. In this case we denote by $<$ the linear order on the arc A such that $a < b$. We will call the order $<$ the *counterclockwise order on A*. Note that $[a, b] \neq [b, a]$.

More generally, for any arc $A = [a, b] \subset \mathbb{S}^1$, with $a < b$ in the counterclockwise order, define the *fractional index* [**Bro90**] of f on the sub-path $g|_{[a,b]}$ by
$$\text{ind}(f, g|_{[a,b]}) = \hat{v}(b) - \hat{v}(a).$$

While, necessarily, the index of f with respect to g is an integer, the fractional index of f on $g|_{[a,b]}$ need not be. We shall have occasion to use fractional index in the proof of Theorem 3.2.2.

PROPOSITION 2.1.2. Let $g : \mathbb{S}^1 \to \mathbb{C}$ be a map with $g(\mathbb{S}^1) = S$, and suppose $f : S \to \mathbb{C}$ has no fixed points on S. Let $a \neq b \in \mathbb{S}^1$ with $[a,b]$ denoting the counterclockwise subarc on $\mathbb{S}^1$ from a to b (so $[a,b]$ and (b,a) are complementary arcs and $\mathbb{S}^1 = [a,b] \cup [b,a]$). Then $\text{ind}(f,g) = \text{ind}(f,g|_{[a,b]}) + \text{ind}(f,g|_{[b,a]})$.

2.2. Variation

In this section we introduce the notion of variation of a map on an arc and relate it to winding number.

DEFINITION 2.2.1 (Junctions). The *standard junction* J_O is the union of the three rays $J_O^i = \{z \in \mathbb{C} \mid z = re^{i\pi/2}, r \in [0,\infty)\}$, $J_O^+ = \{z \in \mathbb{C} \mid z = r, r \in [0,\infty)\}$, $J_O^- = \{z \in \mathbb{C} \mid z - re^{i\pi}, r \in [0,\infty)\}$, having the origin O in common. A *junction (at v)* J_v is the image of J_O under any orientation-preserving homeomorphism $h : \mathbb{C} \to \mathbb{C}$ where $v = h(O)$. We will often suppress h and refer to $h(J_O^i)$ as J_v^i, and similarly for the remaining rays in J_v. Moreover, we require that for each bounded neighborhood W of v, $d(J_v^+ \setminus W, J_v^i \setminus W) > 0$.

DEFINITION 2.2.2 (Variation on an arc). Let $S \subset \mathbb{C}$ be a simple closed curve, $f : S \to \mathbb{C}$ a map and $A = [a,b]$ a subarc of S such that $f(a), f(b) \in T(S)$ and $f(A) \cap A = \emptyset$. We define the *variation of f on A with respect to S*, denoted $\text{var}(f,A,S)$, by the following algorithm:
  (1) Let $v \in A$ and let J_v be a junction with $J_v \cap S = \{v\}$.
  (2) *Counting crossings:* Consider the set $M = f^{-1}(J_v) \cap [a,b]$. Each time a point of $f^{-1}(J_v^+) \cap [a,b]$ is immediately followed in M, in the counter-clockwise order $<$ on $[a,b] \subset S$, by a point of $f^{-1}(J_v^i)$, count $+1$ and each time a point of $f^{-1}(J_v^i) \cap [a,b]$ is immediately followed in M by a point of $f^{-1}(J_v^+)$, count -1. Count no other crossings.
  (3) The sum of the crossings found above is the variation $\text{var}(f,A,S)$.

Note that $f^{-1}(J_v^+) \cap [a,b]$ and $f^{-1}(J_v^i) \cap [a,b]$ are disjoint closed sets in $[a,b]$. Hence, in (2) in the above definition, we count only a finite number of crossings and $\text{var}(f,A,S)$ is an integer. Of course, if $f(A)$ does not meet both J_v^+ and J_v^i, then $\text{var}(f,A,S) = 0$.

If $\alpha : S \to \mathbb{C}$ is any map such that $\alpha|_A = f|_A$ and $\alpha(S \setminus (a,b)) \cap J_v = \emptyset$, then $\text{var}(f,A,S) = \text{win}(\alpha,S,v)$. In particular, this condition is satisfied if $\alpha(S \setminus (a,b)) \subset T(S) \setminus \{v\}$. The invariance of winding number under suitable homotopies implies that the variation $\text{var}(f,A,S)$ also remains invariant under such homotopies. That is, even though the specific crossings in (2) in the algorithm may change, the sum remains invariant. We will state the required results about variation below without proof. Proofs can be obtained directly by using the fact that $\text{var}(f,A,S)$ is integer-valued and continuous under suitable homotopies.

PROPOSITION 2.2.3 (Junction Straightening). Let $S \subset \mathbb{C}$ be a simple closed curve, $f : S \to \mathbb{C}$ a map and $A = [a,b]$ a subarc of S such that $f(a), f(b) \in T(S)$ and $f(A) \cap A = \emptyset$. Any two junctions J_v and J_u with $u, v \in A$ and $J_w \cap S = \{w\}$ for $w \in \{u,v\}$ give the same value for $\text{var}(f,A,S)$. Hence $\text{var}(f,A,S)$ is independent of the particular junction used in Definition 2.2.2.

The computation of $\mathrm{var}(f, A, S)$ depends only upon the crossings of the junction J_v coming from a proper compact subarc of the open arc (a, b). Consequently, $\mathrm{var}(f, A, S)$ remains invariant under homotopies h_t of $f|_{[a,b]}$ in the complement of $\{v\}$ such that $h_t(a), h_t(b) \notin J_v$ for all t. Moreover, the computation is stable under an isotopy $h_t : J_v \to A \cup [\mathbb{C} \setminus T(S)]$ that moves the entire junction J_v (even off A), provided that during the isotopy $h_t(v) \notin f(A)$ and $f(a), f(b) \notin h_t(J_v)$ for all t.

In case A is an open arc $(a, b) \subset S$ such that $\mathrm{var}(f, \overline{A}, S)$ is defined, it will be convenient to denote $\mathrm{var}(f, \overline{A}, S)$ by $\mathrm{var}(f, A, S)$

The following lemma follows immediately from the definition.

LEMMA 2.2.4. *Let $S \subset \mathbb{C}$ be a simple closed curve. Suppose that $a < c < b$ are three points in S such that $\{f(a), f(b), f(c)\} \subset T(S)$ and $f([a, b]) \cap [a, b] = \emptyset$. Then $\mathrm{var}(f, [a, b], S) = \mathrm{var}(f, [a, c], S) + \mathrm{var}(f, [c, b], S)$.*

DEFINITION 2.2.5 (Variation on a finite union of arcs). Let $S \subset \mathbb{C}$ be a simple closed curve and $A = [a, b]$ a subcontinuum of S partitioned by a finite set $F = \{a = a_0 < a_1 < \cdots < a_n = b\}$ into subarcs. For each i let $A_i = [a_i, a_{i+1}]$. Suppose that f satisfies $f(a_i) \in T(S)$ and $f(A_i) \cap A_i = \emptyset$ for each i. We define the *variation of f on A with respect to S*, denoted $\mathrm{var}(f, A, S)$, by

$$\mathrm{var}(f, A, S) = \sum_{i=0}^{n-1} \mathrm{var}(f, [a_i, a_{i+1}], S).$$

In particular, we include the possibility that $a_n = a_0$ in which case $A = S$.

By considering a common refinement of two partitions F_1 and F_2 of an arc $A \subset S$ such that $f(F_1) \cup f(F_2) \subset T(S)$ and satisfying the conditions in Definition 2.2.5, it follows from Lemma 2.2.4 that we get the same value for $\mathrm{var}(f, A, S)$ whether we use the partition F_1 or the partition F_2. Hence, $\mathrm{var}(f, A, S)$ is well-defined. If $A = S$ we denote $\mathrm{var}(f, S, S)$ simply by $\mathrm{var}(f, S)$.

The first main result in Part 1, Theorem 3.2.2 is that given a map $f : \mathbb{C} \to \mathbb{C}$, a simple closed curve $S \subset \mathbb{C}$ and a partition of S into subarcs A_i such that any two meet at most in a common endpoint, for each i $f(A_i) \cap A_i = \emptyset$ and both endpoints map into $T(S)$,

$$\mathrm{ind}(f, S) = \sum \mathrm{var}(f, A_i) + 1.$$

In the first version of this theorem we partition S into finitely many subarcs A_i. We extend this in Section 3.5 by allowing partitions of S which consist of, possibly countably infinitely many subarcs. Since in our applications we often assume that we have an invariant continuum X such that f has no fixed point in $T(X)$ it follows from Theorem 3.1.4 that, for a sufficiently tight simple closed curve S around X with $X \subset T(S)$, we must have $\mathrm{ind}(f, S) = 0$. It follows from the above theorem relating index and variation that for some subarc A (which is the closure of a component of $S \setminus X$ and, hence a crosscut of X), $\mathrm{var}(f, A) < 0$. In order to locate this crosscut of negative variation we establish Bell's Lollipop Theorem in Section 3.3.

2.3. Classes of maps

Cartwright and Littlewood solved the Plane Fixed Point Problem for orientation preserving *homeomorphism* of the plane. In Section 3.7 we introduce and study (positively) oriented *maps* of the plane. We will show in Part 2 that the

Plane Fixed Point Problem has a positive solution for the class of positively oriented maps. We show in Section 3.7 that the class of oriented maps consists of all compositions of monotone and open perfect maps of the plane and that all such maps are confluent. In particular, analytic maps are confluent.

Let us begin by listing a few well-known definitions.

DEFINITION 2.3.1. A *perfect map* is a closed continuous function each of whose point inverses is compact. *We will assume in the remaining sections that all maps of the plane considered in this memoir are perfect.* Let X and Y be topological spaces. A map $f : X \to Y$ is *monotone* provided for each continuum $K \subset Y$, $f^{-1}(K)$ is connected and f is *light* provided for each point $y \in Y$, $f^{-1}(y)$ is totally disconnected. A map $f : X \to Y$ is *confluent* provided for each continuum $K \subset Y$ and each component C of $f^{-1}(K)$, $f(C) = K$. Every map $f : X \to Y$ between compacta is the composition $f = l \circ m$ of a a monotone map $m : X \to Z$ and a light map $l : Z \to Y$ for some compactum Z [**Nad92**, Theorem 13.3]. This representation is called the *monotone-light decomposition of f*.

Observe that any confluent map f is onto. It is well-known that each homeomorphism of the plane is either orientation-preserving or orientation-reversing. We will establish an appropriate extension of this result for confluent perfect mappings of the plane (Theorem 3.7.4) by showing that such maps either preserve or reverse local orientation. As a consequence it follows that all perfect and confluent maps of the plane satisfy the Maximum Modulus Theorem. We will call such maps *positively-* or *negatively oriented* maps, respectively.

Complex polynomials $P : \mathbb{C} \to \mathbb{C}$ are prototypes of positively oriented maps, but positively oriented maps, unlike polynomials, do not have to be light or open. Observe that even though in some applications our maps are holomorphic (see Section 7.5), the notion of a positively oriented map is essential in Section 7.5 since it allows for easy local perturbations (see Lemma 7.5.1).

DEFINITION 2.3.2 (Degree of f_p). Let $f : U \to \mathbb{C}$ be a map from a simply connected domain $U \subset \mathbb{C}$ into the plane. Let $S \subset \mathbb{C}$ be a positively oriented simple closed curve in U, and $p \in U \setminus f^{-1}(f(S))$ a point. Define $f_p : S \to \mathbb{S}^1$ by

$$f_p(x) = \frac{f(x) - f(p)}{|f(x) - f(p)|}.$$

Then f_p has a well-defined *degree*, denoted degree(f_p). Note that degree(f_p) is the winding number win$(f, S, f(p))$ of $f|_S$ about $f(p)$.

DEFINITION 2.3.3. A map $f : U \to \mathbb{C}$ from a simply connected domain U is *positively oriented* (respectively, *negatively oriented*) provided for each simple closed curve S in U and each point $p \in T(S) \setminus f^{-1}(f(S))$, we have that degree($f_p$) > 0 (degree($f_p$) < 0, respectively).

DEFINITION 2.3.4. A perfect surjection $f : \mathbb{C} \to \mathbb{C}$ is *oriented* provided for each simple closed curve S and each $x \in T(S)$, $f(x) \in T(f(S))$.

Clearly every positively oriented and each negatively oriented map is oriented. It will follow that all oriented maps satisfy the Maximum Modulus Theorem 3.7.4 (i.e., for every non-separating continuum X, $\partial f(X) \subset f(\partial X)$). In particular, every positively or negatively oriented map is oriented.

It is well-known that both open maps and monotone maps (and hence compositions of such maps) of continua are confluent. It will follow (Lemma 3.7.3) from a result of Lelek and Read [**LR74**] that each perfect, oriented surjection of the plane is the composition of a monotone map and a light open map.

2.4. Partitioning domains

In Chapter 4 we consider partitions of an open and connected subset U of the sphere into convex subsets which are contained in round balls. Bell originally did this, using the Euclidean metric on the plane, for the complement of X in its convex hull in the plane. Following Kulkarni and Pinkall [**KP94**], we will consider U as a subset of the sphere and we will work with maximal round balls $B \subset U$ in the spherical metric (such balls correspond to either round balls in the plane or to half planes). We first specify Kulkarni and Pinkall's result for our situation (see Theorem 4.1.5). It leads to a partition of U into pairwise disjoint closed subsets F_α such for each α there exists a unique maximal closed round ball B_α with $\text{int}(B_\alpha) \cap \partial U = \emptyset$, $|B_\alpha \cap \partial U| \geq 2$ and $F_\alpha \subset B_\alpha$. In fact, F_α is the intersection of U with the hyperbolic convex hull of $B_\alpha \cap \partial U$ in *the hyperbolic metric on the ball B_α*. Note that every chord in the boundary of any partition element F_α is part of a round circle. This is the partition of U which is used in Part 2, Chapter 6. We show in Section 4.4 that the collection of all chords in the boundaries of all the sets F_α, called $\mathcal{KP}$-*chords*, is sufficiently rich for a satisfactory prime end theory. (Basically most prime ends can be defined through equivalence classes of crosscuts which are all $\mathcal{KP}$-chords.)

However, even though in the above version of the Linchpin Theorem elements of the partition are closed in U and pairwise disjoint and U is an arbitrary connected open subset of the sphere, it has the disadvantage that chords in the boundary of the sets F_α are not naturally depending on U (they depend only on B_α, $B_\alpha \cap \partial U$ and the hyperbolic metric in B_α). Moreover there may well be uncountably many distinct elements F_α which join the same two accessible points in ∂U. In order to avoid this problem we replace, when U is simply connected, any chord in the boundary of any set F_α by the hyperbolic geodesic (*in the hyperbolic metric on U*) joining the same pair of points (see Theorem 4.2.5). We will show that the resulting set of hyperbolic geodesics is a closed lamination of U in the sense of Thurston [**Thu09**]. This version of the Linchpin Theorem, which states that every point in U is either contained in a unique hyperbolic geodesic $\mathfrak{g}$ in U, or in the interior of an unique hyperbolically convex gap $\mathfrak{G}$, both of which are contained in a maximal round ball, is used in [**OT07**, **OV09**] to extend a homeomorphism on the boundary of a simply connected domain over the entire domain. To illustrate the usefulness of these partitions we include a simple proof of the Schoenflies Theorem in Section 4.3. However, we will assume the Schoenflies Theorem throughout this paper.

CHAPTER 3

Tools

3.1. Stability of Index

Let $f : \mathbb{C} \to \mathbb{C}$ be a map. All basic definitions of index of f on a simple closed curve and variation of f on an arc are contained in Chapter 2. The following standard theorems and observations about the stability of index under a fixed point free homotopy are consequences of the fact that index is continuous and integer-valued.

THEOREM 3.1.1. Let $h_t : \mathbb{S}^1 \to \mathbb{C}$ be a homotopy. If $f : \cup_{t \in [0,1]} h_t(\mathbb{S}^1) \to \mathbb{C}$ is fixed point free, then $\operatorname{ind}(f, h_0) = \operatorname{ind}(f, h_1)$.

An embedding $g : \mathbb{S}^1 \to S \subset \mathbb{C}$ is *orientation preserving* if g is isotopic to the identity map $id|_{\mathbb{S}^1}$. It follows from Theorem 3.1.1 that if $g_1, g_2 : \mathbb{S}^1 \to S$ are orientation preserving homeomorphisms and $f : S \to \mathbb{C}$ is a fixed point free map, then $\operatorname{ind}(f, g_1) = \operatorname{ind}(f, g_2)$. Hence we can denote $\operatorname{ind}(f, g_1)$ by $\operatorname{ind}(f, S)$ and if $[a, b]$ is a positively oriented subarc of $\mathbb{S}^1$ we denote the fractional index $\operatorname{ind}(f, g_1|_{[a,b]})$ by $\operatorname{ind}(f, g_1([a, b]))$, by some abuse of notation when the extension of g_1 over $\mathbb{S}^1$ is understood.

THEOREM 3.1.2. Suppose $g : \mathbb{S}^1 \to \mathbb{C}$ is a map with $g(\mathbb{S}^1) = S$, and $f_1, f_2 : S \to \mathbb{C}$ are homotopic maps such that each level of the homotopy is fixed point free on S. Then $\operatorname{ind}(f_1, g) = \operatorname{ind}(f_2, g)$.

In particular, if S is a simple closed curve and $f_1, f_2 : S \to \mathbb{C}$ are maps such that there is a homotopy $h_t : S \to \mathbb{C}$ from f_1 to f_2 with h_t fixed point free on S for each $t \in [0, 1]$, then $\operatorname{ind}(f_1, S) = \operatorname{ind}(f_2, S)$.

COROLLARY 3.1.3. Suppose $g : \mathbb{S}^1 \to \mathbb{C}$ is an orientation preserving embedding with $g(\mathbb{S}^1) = S$, and $f : S \to T(S)$ is a fixed point free map. Then $\operatorname{ind}(f, g) = \operatorname{ind}(f, S) = 1$.

PROOF. Since $f(S) \subset T(S)$ which is a disk with boundary S and f has no fixed point on S, there is a fixed point free homotopy of $f|_S$ to a constant map $c : S \to \mathbb{C}$ taking S to a point in $T(S) \setminus S$. By Theorem 3.1.2, $\operatorname{ind}(f, g) = \operatorname{ind}(c, g)$. Since g is orientation preserving it follows from Remark 2.1.1 (b) that $\operatorname{ind}(c, g) = 1$. □

THEOREM 3.1.4. Suppose $g : \mathbb{S}^1 \to \mathbb{C}$ is a map with $g(\mathbb{S}^1) = S$, and $f : T(S) \to \mathbb{C}$ is a map such that $\operatorname{ind}(f, g) \neq 0$, then f has a fixed point in $T(S)$.

PROOF. Notice that $T(S)$ is a locally connected, non-separating, plane continuum and, hence, contractible. Suppose f has no fixed point in $T(S)$. Choose point $q \in T(S)$. Let $c : \mathbb{S}^1 \to \mathbb{C}$ be the constant map $c(\mathbb{S}^1) = \{q\}$. Let H be a homotopy from g to c with image in $T(S)$. Since H misses the fixed point set of f, Theorem 3.1.1 and Remark 2.1.1 (a) imply $\operatorname{ind}(f, g) = \operatorname{ind}(f, c) = 0$. □

3.2. Index and variation for finite partitions

What links Theorem 3.1.4 with variation is Theorem 3.2.2 below, first announced by Bell in the mid 1980's (see also Akis [**Aki99**]). Our proof is a modification of Bell's unpublished proof. We first need a variant of Proposition 2.2.3. Let $r : \mathbb{C} \to T(\mathbb{S}^1)$ be radial retraction: $r(z) = \frac{z}{|z|}$ when $|z| \geq 1$ and $r|_{T(\mathbb{S}^1)} = id|_{T(\mathbb{S}^1)}$.

LEMMA 3.2.1 (Curve Straightening). Suppose $f : \mathbb{S}^1 \to \mathbb{C}$ is a map with no fixed points on $\mathbb{S}^1$. If $[a, b] \subset \mathbb{S}^1$ is a proper subarc with $f([a, b]) \cap [a, b] = \emptyset$, $f((a, b)) \subset \mathbb{C} \setminus T(\mathbb{S}^1)$ and $f(\{a, b\}) \subset \mathbb{S}^1$, then there exists a map $\tilde{f} : \mathbb{S}^1 \to \mathbb{C}$ such that $\tilde{f}|_{\mathbb{S}^1 \setminus (a,b)} = f|_{\mathbb{S}^1 \setminus (a,b)}$, $\tilde{f}|_{[a,b]} : [a, b] \to (\mathbb{C} \setminus T(\mathbb{S}^1)) \cup \{f(a), f(b)\}$ and $\tilde{f}|_{[a,b]}$ is homotopic to $f|_{[a,b]}$ in $\{a, b\} \cup \mathbb{C} \setminus T(S)$ relative to $\{a, b\}$, so that either $r|_{\tilde{f}([a,b])}$ is locally one-to-one or a constant map. Moreover, $\mathrm{var}(f, [a, b], \mathbb{S}^1) = \mathrm{var}(\tilde{f}, [a, b], \mathbb{S}^1)$.

Note that if $\mathrm{var}(f, [a, b], \mathbb{S}^1) = 0$, then r carries $\tilde{f}([a, b])$ one-to-one onto the arc (or point) in $\mathbb{S}^1 \setminus (a, b)$ from $f(a)$ to $f(b)$. If the $\mathrm{var}(f, [a, b], \mathbb{S}^1) = m > 0$, then $r \circ \tilde{f}$ wraps the arc $[a, b]$ counterclockwise about $\mathbb{S}^1$ so that $\tilde{f}([a, b])$ meets each ray in J_v m times. A similar statement holds for negative variation. Note also that it is possible for index to be defined yet variation not to be defined on a simple closed curve S. For example, consider the map $z \to 2z$ with S the unit circle since there is no partition of S satisfying the conditions in Definition 2.2.2.

THEOREM 3.2.2 (Index = Variation + 1, Bell). Suppose $g : \mathbb{S}^1 \to \mathbb{C}$ is an orientation preserving embedding onto a simple closed curve S and $f : S \to \mathbb{C}$ is a fixed point free map. If $F = \{a_0 < a_1 < \cdots < a_n\}$ is a partition of S and $A_i = [a_i, a_{i+1}]$ for $i = 0, 1, \ldots, n$ with $a_{n+1} = a_0$ such that $f(F) \subset T(S)$ and $f(A_i) \cap A_i = \emptyset$ for each i, then

$$\mathrm{ind}(f, S) = \mathrm{ind}(f, g) = \sum_{i=0}^{n} \mathrm{var}(f, A_i, S) + 1 = \mathrm{var}(f, S) + 1.$$

PROOF. By an appropriate conjugation of f and g, we may assume without loss of generality that $S = \mathbb{S}^1$ and $g = id$. Let F and $A_i = [a_i, a_{i+1}]$ be as in the hypothesis. Consider the collection of arcs

$$\mathcal{K} = \{K \subset S \mid K \text{ is the closure of a component of } f^{-1}(f(S) \setminus T(S))\}.$$

For each $K \in \mathcal{K}$, there is an i such that $K \subset A_i$. Since $f(A_i) \cap A_i = \emptyset$, it follows from the remark after Definition 2.2.2 that $\mathrm{var}(f, A_i, S) = \sum_{K \subset A_i, K \in \mathcal{K}} \mathrm{var}(f, K, S)$. By the remark following Proposition 2.2.3, we can compute $\mathrm{var}(f, K, S)$ using one fixed junction for A_i. It is now clear that there are at most finitely many $K \in \mathcal{K}$ with $\mathrm{var}(f, K, S) \neq 0$. Moreover, the images of the endpoints of each K lie on S.

Let m be the cardinality of the set $\mathcal{K}_f = \{K \in \mathcal{K} \mid \mathrm{var}(f, K, S) \neq 0\}$. By the above remarks, $m < \infty$ and $\mathcal{K}_f$ is independent of the partition F. We prove the theorem by induction on m.

Suppose for a given f we have $m = 0$. Observe that from the definition of variation and the fact that the computation of variation is independent of the choice of an appropriate partition, it follows that,

$$\mathrm{var}(f, S) = \sum_{K \in \mathcal{K}} \mathrm{var}(f, K, S) = 0.$$

3.2. INDEX AND VARIATION FOR FINITE PARTITIONS

We claim that there is a map $f_1 : S \to \mathbb{C}$ with $f_1(S) \subset T(S)$ and a homotopy H from $f|_S$ to f_1 such that each level H_t of the homotopy is fixed point free and $\text{ind}(f_1, id|_S) = 1$.

To see the claim, first apply the Curve Straightening Lemma 3.2.1 to each $K \in \mathcal{K}$ (if there are infinitely many, they form a null sequence) to obtain a fixed point free homotopy of $f|_S$ to a map $\tilde{f} : S \to \mathbb{C}$ such that $r|_{\tilde{f}(K)}$ is locally one-to-one (or the constant map) on each $K \in \mathcal{K}$, where r is radial retraction of $\mathbb{C}$ to $T(S)$, and $\text{var}(\tilde{f}, K, S) = 0$ for each $K \in \mathcal{K}$. Let K be in $\mathcal{K}$ with endpoints x, y. Since $\tilde{f}(K) \cap K = \emptyset$ and $\text{var}(\tilde{f}, K, S) = 0$, $r|_{\tilde{f}(K)}$ is one-to-one, and $r \circ \tilde{f}(K) \cap K = \emptyset$.

Define $f_1|_K = r \circ \tilde{f}|_K$. Then $f_1|_K$ is fixed point free homotopic to $f|_K$ (with endpoints of K fixed). Hence, if $K \in \mathcal{K}$ has endpoints x and y, then f_1 maps K to the subarc of S with endpoints $f(x)$ and $f(y)$ such that $K \cap f_1(K) = \emptyset$. Since $\mathcal{K}$ is a null family, we can do this for each $K \in \mathcal{K}$ and set $f_1|_{\mathbb{S}^1 \setminus \cup \mathcal{K}} = f|_{\mathbb{S}^1 \setminus \cup \mathcal{K}}$ so that we obtain the desired $f_1 : S \to \mathbb{C}$ as the end map of a fixed point free homotopy from f to f_1. Since f_1 carries S into $T(S)$, Corollary 3.1.3 implies $\text{ind}(f_1, id|_S) = 1$.

Since the homotopy $f \simeq f_1$ is fixed point free, it follows from Theorem 3.1.2 that $\text{ind}(f, id|_S) = 1$. Hence, the theorem holds if $m = 0$ for any f and any appropriate partition F.

By way of contradiction suppose the collection $\mathcal{F}$ of all maps f on $\mathbb{S}^1$ which satisfy the hypotheses of the theorem, but not the conclusion is non-empty. By the above $0 < |\mathcal{K}_f| < \infty$ for each. Let $f \in \mathcal{F}$ be a counterexample for which $m = |\mathcal{K}_f|$ is minimal. By modifying f, we will show there exists $f_1 \in \mathcal{F}$ with $|\mathcal{K}_{f_1}| < m$, a contradiction.

Choose $K \in \mathcal{K}$ such that $\text{var}(f, K, S) \neq 0$. Then $K = [x, y] \subset A_i = [a_i, a_{i+1}]$ for some i. By the Curve Straightening Lemma 3.2.1 and Theorem 3.1.2, we may suppose $r|_{f(K)}$ is locally one-to-one on K. Define a new map $f_1 : S \to \mathbb{C}$ by setting $f_1|_{\overline{S \setminus K}} = f|_{\overline{S \setminus K}}$ and setting $f_1|_K$ equal to the linear map taking $[x, y]$ to the subarc $f(x)$ to $f(y)$ on S missing $[x, y]$. Figure 3.1 (left) shows an example of a (straightened) f restricted to K and the corresponding f_1 restricted to K for a case where $\text{var}(f, K, S) = 1$, while Figure 3.1 (right) shows a case where $\text{var}(f, K, S) = -2$.

Since on $\overline{S \setminus K}$, f and f_1 are the same map, we have

$$\text{var}(f, S \setminus K, S) = \text{var}(f_1, S \setminus K, S).$$

Likewise for the fractional index,

$$\text{ind}(f, S \setminus K) = \text{ind}(f_1, S \setminus K).$$

By definition (refer to the observation we made in the case $m = 0$),

$$\text{var}(f, S) = \text{var}(f, S \setminus K, S) + \text{var}(f, K, S)$$
$$\text{var}(f_1, S) = \text{var}(f_1, S \setminus K, S) + \text{var}(f_1, K, S)$$

and by Proposition 2.1.2,

$$\text{ind}(f, S) = \text{ind}(f, S \setminus K) + \text{ind}(f, K)$$
$$\text{ind}(f_1, S) = \text{ind}(f_1, S \setminus K) + \text{ind}(f_1, K).$$

Consequently,

$$\text{var}(f, S) - \text{var}(f_1, S) = \text{var}(f, K, S) - \text{var}(f_1, K, S)$$

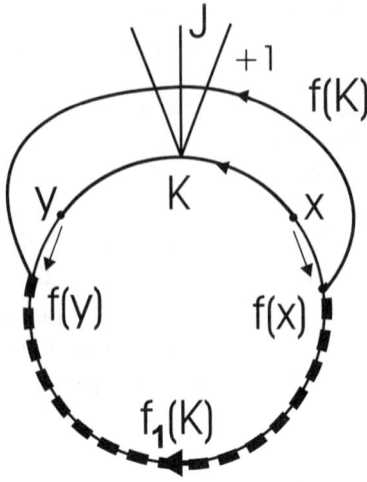 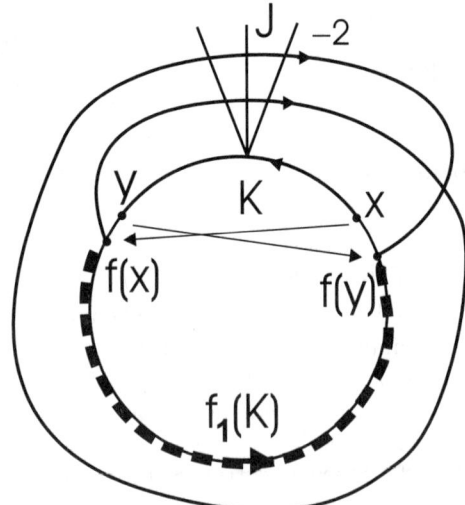

FIGURE 3.1. Replacing $f : S \to \mathbb{C}$ by $f_1 : S \to \mathbb{C}$ with one less subarc of nonzero variation.

and
$$\mathrm{ind}(f, S) - \mathrm{ind}(f_1, S) = \mathrm{ind}(f, K) - \mathrm{ind}(f_1, K).$$

We will now show that the changes in index and variation, going from f to f_1 are the same (i.e., we will show that $\mathrm{var}(f, K, S) - \mathrm{var}(f_1, K, S) = \mathrm{ind}(f, K) - \mathrm{ind}(f_1, K)$). We suppose first that $\mathrm{ind}(f, K) = n + \alpha$ for some nonnegative $n \in \mathbb{N}$ and $0 \le \alpha < 1$. That is, the vector $f(z) - z$ turns through n full revolutions counterclockwise and α part of a revolution counterclockwise as z goes from x to y counterclockwise along S. (See Figure 3.1 (left) for the case $n = 0$ and α about 0.8.)

Assume first that $f(x) < x < y < f(y)$ in the circular order as illustrated in Figure 3.1 on the left. Then as z goes from x to y counterclockwise along S, $f_1(z)$ goes along S from $f(x)$ to $f(y)$ in the clockwise direction, so $f_1(z) - z$ turns through $-(1-\alpha) = \alpha - 1$ part of a revolution. Hence, $\mathrm{ind}(f_1, K) = \alpha - 1$. It is easy to see that $\mathrm{var}(f, K, S) = n + 1$ and $\mathrm{var}(f_1, K, S) = 0$. Consequently,
$$\mathrm{var}(f, K, S) - \mathrm{var}(f_1, K, S) = n + 1 - 0 = n + 1$$
and
$$\mathrm{ind}(f, K) - \mathrm{ind}(f_1, K) = n + \alpha - (\alpha - 1) = n + 1.$$

We assumed that $f(x) < x < y < f(y)$. The cases where $f(y) < x < y < f(x)$ and $f(x) = f(y)$ (and, hence, $\alpha = 0$) are treated similarly. In this case f_1 still wraps around in the positive direction, but the computations are slightly different: $\mathrm{var}(f, K) = n$, $\mathrm{ind}(f, K) = n + \alpha$, $\mathrm{var}(f_1, K) = 0$ and $\mathrm{ind}(f_1, K) = \alpha$.

Thus when $n \ge 0$, in going from f to f_1, the change in variation and the change in index are the same. However, in obtaining f_1 we have removed one $K \in \mathcal{K}_f$, reducing the minimal $m = |\mathcal{K}_f|$ for f by one, producing a counterexample f_1 with $|\mathcal{K}_{f_1}| = m - 1$, a contradiction.

The cases where $\text{ind}(f, K) = n + \alpha$ for negative n and $0 < \alpha < 1$ are handled similarly, and illustrated for $n = -2$, α about 0.4 and $f(y) < x < y < f(x)$ in Figure 3.1 (right). □

3.3. Locating arcs of negative variation

The principal tool in proving Theorem 6.2.1 (unique outchannel) is the following theorem first obtained by Bell (unpublished). It provides a method for locating arcs of negative variation on a curve of index zero.

THEOREM 3.3.1 (Lollipop Lemma, Bell). Let $S \subset \mathbb{C}$ be a simple closed curve and $f : T(S) \to \mathbb{C}$ a fixed point free map. Suppose $F = \{a_0 < \cdots < a_n < a_{n+1} < \cdots < a_m\}$ is a partition of S, $a_{m+1} = a_0$ and $A_i = [a_i, a_{i+1}]$ such that $f(F) \subset T(S)$ and $f(A_i) \cap A_i = \emptyset$ for $i = 0, \ldots, m$. Suppose I is an arc in $T(S)$ meeting S only at its endpoints a_0 and a_{n+1}. Let J_{a_0} be a junction in $(\mathbb{C} \setminus T(S)) \cup \{a_0\}$ and suppose that $f(I) \cap (I \cup J_{a_0}) = \emptyset$. Let $R = T([a_0, a_{n+1}] \cup I)$ and $L = T([a_{n+1}, a_{m+1}] \cup I)$. Then one of the following holds:

(1) If $f(a_{n+1}) \in R$, then
$$\sum_{i \leq n} \text{var}(f, A_i, S) + 1 = \text{ind}(f, I \cup [a_0, a_{n+1}]).$$

(2) If $f(a_{n+1}) \in L$, then
$$\sum_{i > n} \text{var}(f, A_i, S) + 1 = \text{ind}(f, I \cup [a_{n+1}, a_{m+1}]).$$

(Note that in (1) in effect we compute $\text{var}(f, \partial R)$ but technically, we have not defined $\text{var}(f, A_i, \partial R)$ since the endpoints of A_i do not have to map inside R but they do map into $T(S)$. Similarly in Case (2).)

PROOF. Suppose $f(a_{n+1}) \in L$ (the case when $f(a_{n+1}) \in R$ can be treated similarly). Consider the set $C = [a_{n+1}, a_{m+1}] \cup I$ (so $T(C) = L$). We want to construct a map $f' : C \to \mathbb{C}$, fixed point free homotopic to $f|_C$, that does not change variation on any arc A_i in C and has the properties listed below.

(1) $f'(a_i) \in L$, $f'(A_i) \cap A_i = \emptyset$ for all $n+1 \leq i \leq m$ and $f'(a_0) \in L$. Hence $\text{var}(f', A_i, C)$ is defined for each $i > n$.
(2) $\text{var}(f', A_i, C) = \text{var}(f, A_i, S)$ for all $n+1 \leq i \leq m$.
(3) $f'(I) \cap I = \emptyset$ and $\text{var}(f', I, C) = 0$.

Having such a map, it then follows from Theorem 3.2.2, that
$$\text{ind}(f', C) = \sum_{i=n+1}^{m} \text{var}(f', A_i, C) + \text{var}(f', I, C) + 1.$$

By Theorem 3.1.2 $\text{ind}(f', C) = \text{ind}(f, C)$. By (2) and (3), $\sum_{i>n} \text{var}(f', A_i, C) + \text{var}(f', I, C) = \sum_{i>n} \text{var}(f, A_i, S)$ and the Theorem would follow.

It remains to define the map $f' : C \to \mathbb{C}$ with the above properties. For each i such that $n+1 \leq i \leq m+1$, chose an arc I_i joining $f(a_i)$ to L as follows:

(a) If $f(a_i) \in L$, let I_i be the degenerate arc $\{f(a_i)\}$.
(b) If $f(a_i) \in R$ and $n+1 < i < m+1$, let I_i be an arc in $R \setminus \{a_0, a_{n+1}\}$ joining $f(a_i)$ to I.
(c) If $f(a_0) \in R$, let I_0 be an arc joining $f(a_0)$ to L such that $I_0 \cap (L \cup J_{a_0}) \subset A_{n+1} \setminus \{a_{n+1}\}$.

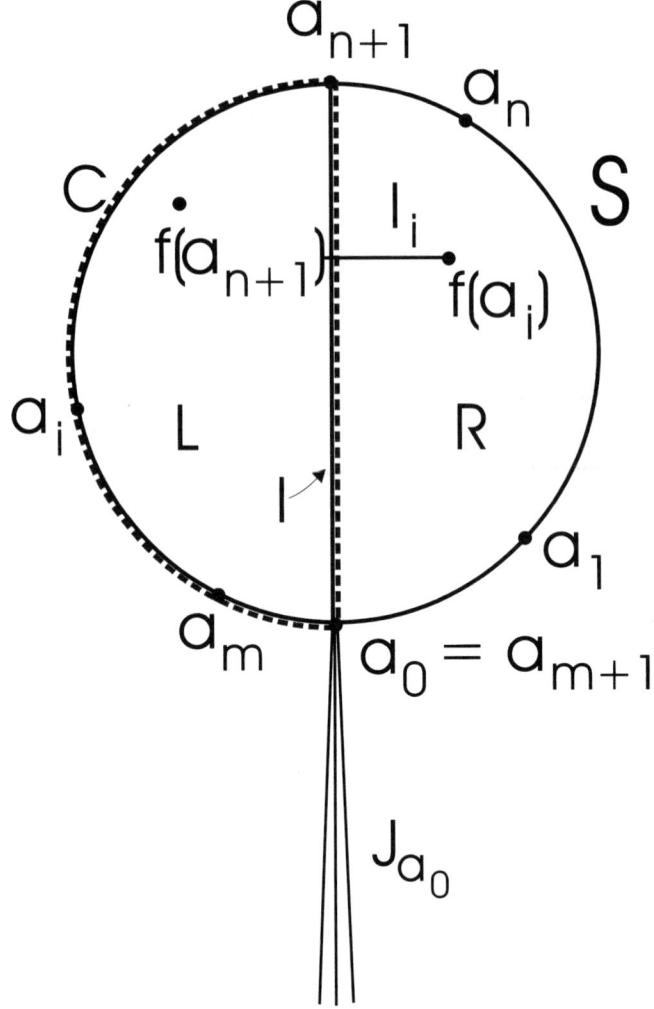

FIGURE 3.2. Bell's Lollipop.

Let $x_{n+1} = y_{n+1} = a_{n+1}$, $y_0 = y_{m+1} \in I \setminus \{a_0, a_{n+1}\}$ and $x_0 = x_{m+1} \in A_m \setminus \{a_m, a_{m+1}\}$. For $n+1 < i < m+1$, let $x_i \in A_{i-1}$ and $y_i \in A_i$ such that $y_{i-1} < x_i < a_i < y_i < x_{i+1}$. For $n+1 < i < m+1$ let $f'(a_i)$ be the endpoint of I_i in L, $f'(x_i) = f'(y_i) = f(a_i)$ and extend f' continuously from $[x_i, a_i] \cup [a_i, y_i]$ onto I_i and define f' from $[y_i, x_{i+1}] \subset A_i$ onto $f(A_i)$ by $f'|_{[y_i, x_{i+1}]} = f \circ h_i$, where $h_i : [y_i, x_{i+1}] \to A_i$ is a homeomorphism such that $h_i(y_i) = a_i$ and $h_i(x_{i+1}) = a_{i+1}$. Similarly, define f' on $[y_0, a_{n+1}] \subset I$ to $f(I)$ by $f'|_{[y_0, a_{n+1}]} = f \circ h_0$, where $h_0 : [y_0, a_{n+1}] \to I$ is an onto homeomorphism such that $h(a_{n+1}) = a_{n+1}$ and extend f' from $[x_{m+1}, a_0] \subset A_m$ and $[a_0, y_0] \subset I$ onto I_0 such that $f'(x_{m+1}) = f'(y_0) = f(a_0)$ and $f'(a_0)$ is the endpoint of I_0 in L. To define $f'|_{[a_{n+1}, x_{n+2}]}$ let $h_{n+1} : [y_{n+1}, x_{n+2}] \to [a_{n+1}, a_{n+2}]$ be a homeomorphism such that $h_{n+1}(y_{n+1}) = a_{n+1}$. Then define $f'(x)$ as $f \circ h_{n+1}(x)$ for $x \in [y_{n+1}, x_{n+2}]$ and $f'(x) = f(a_{n+1})$ if $x \in [a_{n+1}, y_{n+1}]$.

Note that $f'(A_i) \cap A_i = \emptyset$ for $i = n+1, \ldots, m$ and $f'(I) \cap [I \cup J_{a_0}] = \emptyset$. To compute the variation of f' on each of A_m and I we can use the junction J_{a_0}. Hence $\text{var}(f', I, C) = 0$ and, by the definition of f' on A_m, $\text{var}(f', A_m, C) = \text{var}(f, A_m, S)$. For $i = n+1, \ldots, m-1$ we can use the same junction J_{v_i} to compute $\text{var}(f', A_i, C)$ as we did to compute $\text{var}(f, A_i, S)$. Since $I_i \cup I_{i+1} \subset T(S) \setminus A_i$ we have that $f'([a_i, y_i]) \cup f'([x_{i+1}, a_{i+1}]) \subset I_i \cup I_{i+1}$ misses that junction and, hence, make no contribution to variation $\text{var}(f', A_i, C)$. Since $f'^{-1}(J_{v_i}) \cap [y_i, x_{i+1}]$ is isomorphic to $f^{-1}(J_{v_i}) \cap A_i$, $\text{var}(f', A_i, C) = \text{var}(f, A_i, S)$ for $i = n+1, \ldots, m$.

To see that f' is fixed point free homotopic to $f|_C$, note that we can pull the image of A_i back along the arcs I_i and I_{i+1} in R without fixing a point of A_i at any level of the homotopy. □

Note that if f is fixed point free on $T(S)$, then $\text{ind}(f, C) = 0$ and the next Corollary follows.

COROLLARY 3.3.2. *Assume the hypotheses of Theorem 3.3.1. Then if $f(a_{n+1}) \in R$ there exists $i \leq n$ such that $\text{var}(f, A_i, S) < 0$. If $f(a_{n+1}) \in L$ there exists $i > n$ such that $\text{var}(f, A_i, S) < 0$.*

3.4. Crosscuts and bumping arcs

For the remainder of Chapter 3, we assume that $f : \mathbb{C} \to \mathbb{C}$ takes the continuum X into $T(X)$ with no fixed points in $T(X)$, and X is minimal with respect to these properties.

DEFINITION 3.4.1 (Bumping Simple Closed Curve). A simple closed curve S in $\mathbb{C}$ which has the property that $S \cap X$ is nondegenerate and $T(X) \subset T(S)$ is said to be a *bumping simple closed curve for X*. A subarc A of a bumping simple closed curve, whose endpoints lie in X, is said to be a *bumping (sub)arc for X* or a *link of S*. Moreover, if S' is any bumping simple closed curve for X which contains A, then S' is said to *complete A*. In fact, an arc A with endpoints in X which can be completed will be called a *bumping arc of X*.

Given a positively oriented simple closed curve S, we can consider its positively oriented subarcs denoted by $[a, b]_S$, where a, b are the endpoints of the arc; if the curve is fixed, we simply write $[a, b]$. Similar notation is used for half-open or open subarcs of bumping simple closed curves. Often we will fix the choice of links into which we divide S. In general a bumping arc of X may have points other than its endpoints which belong to X (e.g., if X is the closed unit disk and the unit circle is divided into several subarcs then each of them can be considered as a bumping arc of X). By definition, any bumping arc A of X can be extended to a bumping simple closed curve S of X. Hence, every bumping arc has a well defined natural order $<$ inherited from the positive circular order of a bumping simple closed curve S containing A. If $a < b$ are the endpoints of A, then we will often write $A = [a, b]$.

A *crosscut* of $U^\infty = \mathbb{C}^\infty \setminus T(X)$ is an open arc Q lying in $U^\infty \setminus \{\infty\}$ such that $\overline{Q}$ is an arc with endpoints $a \neq b \in T(X)$. In this case we will often write $Q = (a, b)$. (As seems to be traditional, we use "crosscut of $T(X)$" interchangeably with "crosscut of U^∞.") Evidently, a crosscut of U^∞ separates U^∞ into two disjoint domains, exactly one of which is unbounded. If S is a bumping simple closed curve so that $X \cap S$ is nondegenerate, then each component of $S \setminus X$ is a crosscut of $T(X)$. A similar statement holds for a bumping arc A. Given a non-separating continuum $T(X)$, let $A \subset \mathbb{C}$ be a crosscut of $U^\infty(X) = \mathbb{C}^\infty \setminus T(X)$. Given a crosscut A of $U^\infty(X)$ denote by $\mathrm{Sh}(A)$, the *shadow of* A, the bounded component of $\mathbb{C} \setminus [T(X) \cup A]$. Moreover, suppose that A is a bumping arc of X. Then by the *shadow* $\mathrm{Sh}(A)$ *of* A, we mean the union of all bounded components of $\mathbb{C} \setminus (X \cup A)$ (since there may be more than endpoints of A in $X \cap A$, we should talk about the union of all bounded components of $\mathbb{C} \setminus (X \cup A)$ here).

A variety of tools (such as index, variation, junction) have been described in previous sections. So far they have been applied to the properties of maps of the plane restricted to simple closed curves. Another application can be found in Theorem 3.1.4 where the existence of a fixed point in the topological hull of a simple closed curve is established. However we are mostly interested in studying continua X as described in our Standing Hypotheses. The following construction shows how the above described tools apply in this situation.

Since f has no fixed points in $T(X)$ and X is compact, we can choose a bumping simple closed curve S in a small neighborhood of $T(X)$ such that all crosscuts in $S \setminus X$ are small, have positive distance to their image and so that f has no fixed points in $T(S)$. Thus, we obtain the following corollary to Theorem 3.1.4.

COROLLARY 3.4.2. *Let $f : \mathbb{C} \to \mathbb{C}$ be a map and $X \subset \mathbb{C}$ a subcontinuum with $f(X) \subset T(X)$ and so that $f|_{T(X)}$ is fixed point free. Then there is a bumping simple closed curve S for X such that $f|_{T(S)}$ is fixed point free; hence, by 3.1.4, $\mathrm{ind}(f, S) = 0$. Moreover, any bumping simple closed curve S' for X such that $S' \subset T(S)$ has $\mathrm{ind}(f, S') = 0$. Furthermore, any bumping arc A of $T(X)$ for which f has no fixed points in $T(X \cup A)$ can be completed to a bumping simple closed curve S for X for which $\mathrm{ind}(f, S) = 0$.*

The idea of the proofs of a few forthcoming results is that in some cases we can use the developed tools (e.g., variation) in order to compute out index and show, relying upon the properties of our maps, that index is *not* equal to zero thus contradicting Corollary 3.4.2. To implement such a plan we need to further study properties of variation in the setting described before Corollary 3.4.2.

PROPOSITION 3.4.3. *Let $f : \mathbb{C} \to \mathbb{C}$ be a map and $X \subset \mathbb{C}$ a subcontinuum with $f(X) \subset T(X)$ and so that $f|_{T(X)}$ is fixed point free. In the situation of Corollary 3.4.2, suppose A is a bumping subarc for X. If $\mathrm{var}(f, A, S)$ is defined for some bumping simple closed curve S completing A, then for any bumping simple closed curve S' completing A, $\mathrm{var}(f, A, S) = \mathrm{var}(f, A, S')$.*

PROOF. Since $\mathrm{var}(f, A, S)$ is defined, $A = \cup_{i=1}^n A_i$, where each A_i is a bumping arc with $A_i \cap f(A_i) = \emptyset$ and $|A_i \cap A_j| \le 1$ if $i \ne j$. By the remark following Definition 2.2.5, it suffices to establish the desired result for each $A_i = A$. Let S and S' be two bumping simple closed curves completing A for which variation is defined. Let J_a and $J_{a'}$ be junctions whereby $\mathrm{var}(f, A, S)$ and $\mathrm{var}(f, A, S')$ are respectively computed. Suppose first that both junctions lie (except for $\{a, a'\}$) in $\mathbb{C} \setminus (T(S) \cup T(S'))$. By the Junction Straightening Proposition 2.2.3, either junction can be used to compute either variation on A, so the result follows. Otherwise, at least one junction is not in $\mathbb{C} \setminus (T(S) \cup T(S'))$. But both junctions are in $\mathbb{C} \setminus T(X \cup A)$. Hence, we can find another bumping simple closed curve S'' such that S'' completes A, and both junctions lie in $(\mathbb{C} \setminus T(S'')) \cup \{a, a'\}$. Then by the Propositions 2.2.3 and the definition of variation, $\mathrm{var}(f, A, S) = \mathrm{var}(f, A, S'') = \mathrm{var}(f, A, S')$. □

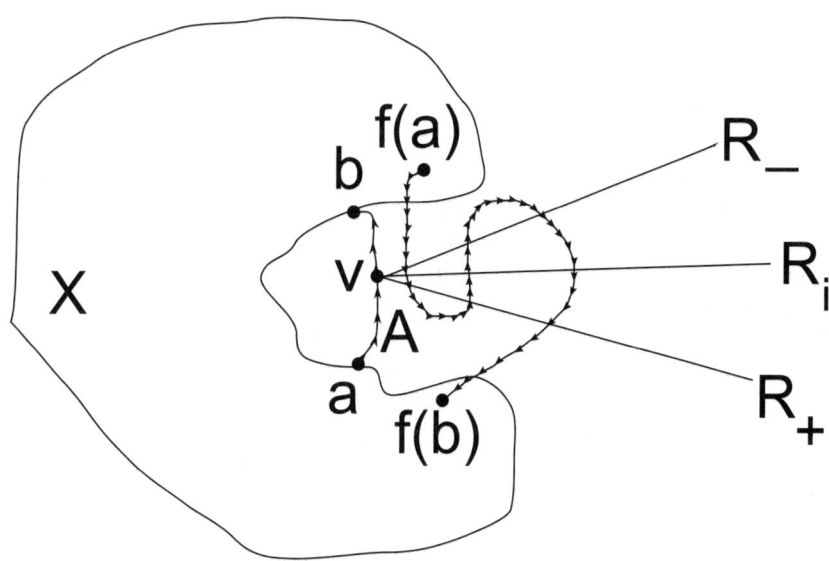

FIGURE 3.3. $\mathrm{var}(f, A) = -1 + 1 - 1 = -1$.

It follows from Proposition 3.4.3 that variation on a crosscut Q, with $\overline{Q} \cap f(\overline{Q}) = \emptyset$, of $T(X)$ is independent of the bumping simple closed curve S for $T(X)$ of which Q is a subarc and is such that $\mathrm{var}(f, Q, S)$ is defined. Hence, given a bumping arc A of X, we can denote $\mathrm{var}(f, A, S)$ by $\mathrm{var}(f, A, X)$ or simply by $\mathrm{var}(f, A)$ when X is understood. The figure illustrates how variation is computed.

The following proposition follows from Corollary 3.4.2, Proposition 3.4.3 and Theorem 3.2.2.

PROPOSITION 3.4.4. Let $f : \mathbb{C} \to \mathbb{C}$ be a map and X a subcontinuum of $\mathbb{C}$ so that $f(X) \subset T(X)$ and f has no fixed points in $T(X)$. Suppose Q is a crosscut of $T(X)$ such that f is fixed point free on $T(X \cup Q)$ and $f(\overline{Q}) \cap \overline{Q} = \emptyset$. Suppose Q is replaced by a bumping subarc A with the same endpoints such that $Q \cup T(X)$ separates $A \setminus X$ from ∞ and each component Q_i of $A \setminus X$ is a crosscut such that $f(\overline{Q_i}) \cap \overline{Q_i} = \emptyset$. Then

$$\mathrm{var}(f, Q, X) = \sum_i \mathrm{var}(f, Q_i, X) = \mathrm{var}(f, A, X).$$

3.5. Index and Variation for Carathéodory Loops

We extend the definitions of index and variation to *Carathéodory loops*. In particular, if $g : \mathbb{S}^1 \to g(\mathbb{S}^1) = S$ is a continuous extension of a Riemann map $\psi : \mathbb{D}^\infty \to \mathbb{C}^\infty \setminus T(g(\mathbb{S}^1))$, then g is a Carathéodory loop, where $\mathbb{D}^\infty = \{z \in \mathbb{C}^\infty \mid |z| > 1\}$ is the "unit disk" about ∞.

DEFINITION 3.5.1 (Carathéodory Loop). Let $g : \mathbb{S}^1 \to \mathbb{C}$ such that g is continuous and has a continuous extension $\psi : \overline{\mathbb{C}^\infty \setminus T(\mathbb{S}^1)} \to \overline{\mathbb{C}^\infty \setminus T(g(\mathbb{S}^1))}$ such that $\psi|_{\mathbb{C}\setminus T(\mathbb{S}^1)}$ is an orientation preserving homeomorphism from $\mathbb{C} \setminus T(\mathbb{S}^1)$ onto $\mathbb{C} \setminus T(g(\mathbb{S}^1))$. We call g (and loosely, $S = g(\mathbb{S}^1)$), a *Carathéodory loop*.

Let $g : \mathbb{S}^1 \to \mathbb{C}$ be a Carathéodory loop and let $f : g(\mathbb{S}^1) \to \mathbb{C}$ be a fixed point free map. In order to define variation of f on $g(\mathbb{S}^1)$, we do the partitioning in $\mathbb{S}^1$ and transport it to the Carathéodory loop $S = g(\mathbb{S}^1)$. An *allowable* partition of $\mathbb{S}^1$ is a set $\{a_0 < a_1 < \cdots < a_n\}$ in $\mathbb{S}^1$ ordered counterclockwise, where $a_0 = a_n$ and A_i denotes the counterclockwise interval $[a_i, a_{i+1}]$, such that for each i, $f(g(a_i)) \in T(g(\mathbb{S}^1))$ and $f(g(A_i)) \cap g(A_i) = \emptyset$. Variation $\text{var}(f, A_i, g(\mathbb{S}^1)) = \text{var}(f, A_i)$ on each path $g(A_i)$ is then defined exactly as in Definition 2.2.2, except that the junction (see Definition 2.2.1) is chosen so that the vertex $v \in g(A_i)$ and $J_v \cap T(g(\mathbb{S}^1)) \subset \{v\}$, and the crossings of the junction J_v by $f(g(A_i))$ are counted (see Definition 2.2.2). Variation on the whole loop, or an allowable subarc thereof, is defined just as in Definition 2.2.5, by adding the variations on the partition elements. At this point in the development, variation is defined only relative to the given allowable partition F of $\mathbb{S}^1$ and the parameterization g of S: $\text{var}(f, F, g(\mathbb{S}^1))$.

Index on a Carathéodory loop S is defined exactly as in Section 2.1 with $S = g(\mathbb{S}^1)$ providing the parameterization of S. Likewise, the definition of fractional index and Proposition 2.1.2 apply to Carathéodory loops.

Theorems 3.1.1, 3.1.2, Corollary 3.1.3, and Theorem 3.1.4 (if f is also defined on $T(S)$) apply to Carathéodory loops. It follows that index on a Carathéodory loop S is independent of the choice of parameterization g. The Carathéodory loop S is approximated, under small homotopies, by simple closed curves S_i. Allowable partitions of S can be made to correspond to allowable partitions of S_i under small homotopies. Since variation and index are invariant under suitable homotopies (see the comments after Proposition 2.2.3) we have the following theorem.

THEOREM 3.5.2. *Suppose $S = g(\mathbb{S}^1)$ is a parameterized Carathéodory loop in $\mathbb{C}$ and $f : S \to \mathbb{C}$ is a fixed point free map. Suppose variation of f on $\mathbb{S}^1 = A_0 \cup \cdots \cup A_n$ with respect to g is defined for some partition $A_0 \cup \cdots \cup A_n$ of $\mathbb{S}^1$. Then*

$$\text{ind}(f, g) = \sum_{i=0}^{n} \text{var}(f, A_i, g(\mathbb{S}^1)) + 1.$$

3.6. Prime Ends

Prime ends provide a way of studying the approaches to the boundary of a simply-connected plane domain with non-degenerate boundary. See [**CL66**] or [**Mil00**] for an analytic summary of the topic and [**UY51**] for a more topological approach. We will be interested in the prime ends of $U^\infty = \mathbb{C}^\infty \setminus T(X)$. Recall that $\mathbb{D}^\infty = \{z \in \mathbb{C}^\infty \mid |z| > 1\}$ is the "unit disk about ∞." The Riemann Mapping Theorem guarantees the existence of a conformal map $\phi : \mathbb{D}^\infty \to U^\infty$ taking $\infty \to \infty$, unique up to the argument of the derivative at ∞. Fix such a map ϕ. We identify $\mathbb{S}^1 = \partial \mathbb{D}^\infty$ with $\mathbb{R}/\mathbb{Z}$ and identify points $e^{2\pi i t}$ in $\partial \mathbb{D}^\infty$ by their argument t (mod 1). Crosscut and shadow were defined in Section 3.4.

DEFINITION 3.6.1 (Prime End). A *chain of crosscuts* is a sequence $\{Q_i\}_{i=1}^{\infty}$ of crosscuts of U^∞ such that for $i \neq j$, $\overline{Q}_i \cap \overline{Q}_j = \emptyset$, $\text{diam}(Q_i) \to 0$, and for all $j > i$, Q_i separates Q_j from ∞ in U^∞. Hence, for all $j > i$, $Q_j \subset \text{Sh}(Q_i)$. Two chains of crosscuts are said to be *equivalent* if and only if it is possible to form a sequence of crosscuts by selecting alternately a crosscut from each chain so that the resulting sequence of crosscuts is again a chain. A *prime end* $\mathcal{E}$ is an equivalence class of chains of crosscuts.

If $\{Q_i\}$ and $\{Q'_i\}$ are equivalent chains of crosscuts of U^∞, it can be shown that $\{\phi^{-1}(Q_i)\}$ and $\{\phi^{-1}(Q'_i)\}$ are equivalent chains of crosscuts of $\mathbb{D}^\infty$ each of which converges to the same unique point $e^{2\pi i t} \in \mathbb{S}^1 = \partial \mathbb{D}^\infty$, $t \in [0,1)$, independent of the representative chain. Hence, we denote by $\mathcal{E}_t$ the prime end of U^∞ defined by $\{Q_i\}$.

DEFINITION 3.6.2 (Impression and Principal Continuum). Let $\mathcal{E}_t$ be a prime end of U^∞ with defining chain of crosscuts $\{Q_i\}$. The set

$$\mathrm{Im}(\mathcal{E}_t) = \bigcap_{i=1}^{\infty} \overline{\mathrm{Sh}(Q_i)}$$

is a subcontinuum of ∂U^∞ called the *impression* of $\mathcal{E}_t$. The set

$$\mathrm{Pr}(\mathcal{E}_t) = \{z \in \partial U^\infty \mid \text{for some chain } \{Q'_i\} \text{ defining } \mathcal{E}_t,\, Q'_i \to z\}$$

is a continuum called the *principal continuum* of $\mathcal{E}_t$.

For a prime end $\mathcal{E}_t$, $\mathrm{Pr}(\mathcal{E}_t) \subset \mathrm{Im}(\mathcal{E}_t)$, possibly properly. We will be interested in the existence of prime ends $\mathcal{E}_t$ for which $\mathrm{Pr}(\mathcal{E}_t) = \mathrm{Im}(\mathcal{E}_t) = \partial U^\infty$.

DEFINITION 3.6.3 (External Rays). Let $t \in [0,1)$ and define

$$R_t = \{z \in \mathbb{C} \mid z = \phi(re^{2\pi i t}), 1 < r < \infty\}.$$

We call R_t the *external ray (with argument t)*. If $x \in R_t$ then the (X,x)-*end* of R_t is the bounded component K_x of $R_t \setminus \{x\}$.

In this case X is a continuum, $U^\infty(X)$ is simply connected, the external rays R_t are all smooth and pairwise disjoint. Moreover, for each $x \in U^\infty(X)$ there exists a unique t such that $x \in R_t$.

DEFINITION 3.6.4 (Essential crossing). An external ray R_t is said to *cross* a crosscut Q *essentially* if and only if there exists $x \in R_t$ such that the $(T(X),x)$-end of R_t is contained in the bounded complementary domain of $T(X) \cup Q$. In this case we will also say that Q crosses R_t essentially.

The results listed below are known.

PROPOSITION 3.6.5 ([**CL66**]). Let $\mathcal{E}_t$ be a prime end of U^∞. Then $\mathrm{Pr}(\mathcal{E}_t) = \overline{R_t} \setminus R_t$. Moreover, for each $1 < r < \infty$ there is a crosscut Q_r of U^∞ with $\{\phi(re^{2\pi i t})\} = R_t \cap Q_r$ and $\mathrm{diam}(Q_r) \to 0$ as $r \to 1$ and such that R_t crosses Q_r essentially.

DEFINITION 3.6.6 (Landing Points and Accessible Points). If $\mathrm{Pr}(\mathcal{E}_t) = \{x\}$, then we say R_t *lands on* $x \in T(X)$ and x is the *landing point* of R_t. A point $x \in \partial T(X)$ is said to be *accessible* (from U^∞) if and only if there is an arc in $U^\infty \cup \{x\}$ with x as one of its endpoints.

PROPOSITION 3.6.7. A point $x \in \partial T(X)$ is accessible if and only if x is the landing point of some external ray R_t.

DEFINITION 3.6.8 (Channels). A prime end $\mathcal{E}_t$ of U^∞ for which $\mathrm{Pr}(\mathcal{E}_t)$ is non-degenerate is said to be a *channel in* ∂U^∞ (or in $T(X)$). If moreover $\mathrm{Pr}(\mathcal{E}_t) = \partial U^\infty = \partial T(X)$, we say $\mathcal{E}_t$ is a *dense* channel. A crosscut Q of U^∞ is said to *cross* the channel $\mathcal{E}_t$ if and only if R_t crosses Q essentially.

When X is locally connected, there are no channels, as the following classical theorem proves. In this case, every prime end has degenerate principal set and degenerate impression.

THEOREM 3.6.9 (Carathéodory). X is locally connected if and only if the Riemann map $\phi : \mathbb{D}^\infty \to U^\infty = \mathbb{C}^\infty \setminus T(X)$ taking $\infty \to \infty$ extends continuously to $\mathbb{S}^1 = \partial \mathbb{D}^\infty$.

3.7. Oriented maps

Basic notions of (positively) oriented and confluent maps are defined in Chapter 2. In this section we study (positively) oriented maps and we will establish that it is a natural class of plane maps which are the proper generalization of an orientation preserving homeomorphism of the plane. The following lemmas are in preparation for the proof of Theorem 3.7.4.

LEMMA 3.7.1. Suppose $f : \mathbb{C} \to \mathbb{C}$ is a perfect surjection. Then f is confluent if and only if f is oriented.

PROOF. Suppose that f is oriented. Let A be an arc in $\mathbb{C}$ and let C be a component of $f^{-1}(A)$. Suppose that $f(C) \neq A$. Let $a \in A \setminus f(C)$. Since $f(C)$ does not separate a from infinity, we can choose a simple closed curve S with $C \subset T(S)$, $S \cap f^{-1}(A) = \emptyset$ and $f(S)$ so close to $f(C)$ that $f(S)$ does not separate a from ∞. Then $a \notin T(f(S))$. Since f is oriented, $f(C) \subset T(f(S))$. Hence there exists a $y \in A \cap f(S)$. This contradicts the fact that $A \cap f(S) = \emptyset$. Thus $f(C) = A$.

Now suppose that K is an arbitrary continuum in $\mathbb{C}$ and let L be a component of $f^{-1}(K)$. Let $x \in L$ and let A_i be a sequence of arcs in $\mathbb{C}$ such that $\lim A_i = K$ and $f(x) \in A_i$ for each i. Let M_i be the component of $f^{-1}(A_i)$ containing the point x. By the previous paragraph $f(M_i) = A_i$. Since f is perfect, $M = \limsup M_i \subset L$ is a continuum and $f(M) = K$. Hence f is confluent.

Suppose next that $f : \mathbb{C} \to \mathbb{C}$ is not oriented. Then there exists a simple closed curve S in $\mathbb{C}$ and $p \in T(S)$ such that $f(p) \notin T(f(S))$. Let L be a half-line with endpoint $f(p)$ running to infinity in $\mathbb{C} \setminus f(S)$. Let L^* be an arc in L with endpoint $f(p)$ and diameter greater than the diameter of the continuum $f(T(S))$. Let K be the component of $f^{-1}(L^*)$ which contains p. Then $K \subset T(S)$, since $p \in T(S)$ and $L \cap f(S) = \emptyset$. Hence, $f(K) \neq L^*$, and so f is not confluent. $\square$

LEMMA 3.7.2. Let $f : \mathbb{C} \to \mathbb{C}$ be a light, open, perfect surjection. Then there exists an integer k and a finite subset $B \subset \mathbb{C}$ such that f is a local homeomorphism at each point of $\mathbb{C} \setminus B$, and for each point $y \in \mathbb{C} \setminus f(B)$, $|f^{-1}(y)| = k$.

PROOF. Let $\mathbb{C}^\infty$ be the one point compactification of $\mathbb{C}$. Since f is perfect, we can extend f to a map of $\mathbb{C}^\infty$ onto $\mathbb{C}^\infty$ so that $f^{-1}(\infty) = \infty$. By abuse of notation we also denote the extended map by f. Then f is a light open mapping of the compact 2-manifold $\mathbb{C}^\infty$. The result now follows from a theorem of Whyburn [**Why42**, X.6.3]. $\square$

The following is a special case, for oriented perfect maps, of the monotone-light factorization theorem. A non-separating plane continuum is said to be *acyclic*.

LEMMA 3.7.3. Suppose that $f : \mathbb{C} \to \mathbb{C}$ is an oriented, perfect map. It follows that $f = g \circ h$, where $h : \mathbb{C} \to \mathbb{C}$ is a monotone perfect surjection with acyclic fibers and $g : \mathbb{C} \to \mathbb{C}$ is a light, open perfect surjection.

PROOF. As above, f extends to a map of the sphere such that $f(\infty) = f^{-1}(\infty) = \infty$. By the monotone-light factorization theorem [**Nad92**, Theorem 13.3], $f = g \circ h$, where $h : \mathbb{C} \to X$ is monotone, $g : X \to \mathbb{C}$ is light, and X is the quotient space obtained from $\mathbb{C}$ by identifying each component of $f^{-1}(y)$ to a point for each $y \in \mathbb{C}$. Let $y \in \mathbb{C}$ and let C be a component of $f^{-1}(y)$. If C were to separate $\mathbb{C}$, then $f(C) = y$ would be a point while $f(T(C))$ would be a non-degenerate continuum. Choose an arc $A \subset \mathbb{C} \setminus \{y\}$ which meets both $f(T(C))$ and its complement and let $x \in T(C) \setminus C$ such that $f(x) \in A$. If K is the component of $f^{-1}(A)$ which contains x, then $K \subset T(C)$. Hence $f(K)$ cannot map onto A contradicting the fact that f is confluent. Thus, for each $y \in \mathbb{C}$, each component of $f^{-1}(y)$ is acyclic.

By Moore's Plane Decomposition Theorem [**Dav86**], X is homeomorphic to $\mathbb{C}$. Since f is confluent, it is easy to see that g is confluent. By a theorem of Lelek and Read [**LR74**] g is open since it is confluent and light (also see [**Nad92**, Theorem 13.26]). Since h and g factor the perfect map f through a Hausdorff space $\mathbb{C}$, both h and g are perfect [**Eng89**, 3.7.5]. □

Below ∂Z means the boundary of the set Z.

THEOREM 3.7.4 (Maximum Modulus Theorem). Suppose that $f : \mathbb{C} \to \mathbb{C}$ is a perfect surjection. Then the following are equivalent:

(1) f is either positively or negatively oriented.
(2) f is oriented.
(3) f is confluent.

Moreover, if f is oriented, then for any non-separating continuum X, $\partial f(X) \subset f(\partial X)$.

PROOF. It is clear that (1) implies (2). By Lemma 3.7.1 every oriented map is confluent. Hence suppose that $f : \mathbb{C} \to \mathbb{C}$ is a perfect confluent map. By Lemma 3.7.3, $f = g \circ h$, where $h : \mathbb{C} \to \mathbb{C}$ is a monotone perfect surjection with acyclic fibers and $g : \mathbb{C} \to \mathbb{C}$ is a light, open perfect surjection. By Stoilow's Theorem [**Why64**] there exists a homeomorphism $j : \mathbb{C} \to \mathbb{C}$ such that $g \circ j$ is an analytic surjection. Then $f = g \circ h = (g \circ j) \circ (j^{-1} \circ h)$. Since $k = j^{-1} \circ h$ is a monotone surjection of $\mathbb{C}$ with acyclic fibers, it is a near homeomorphism [**Dav86**, Theorem 25.1]. That is, there exists a sequence k_i of homeomorphisms of $\mathbb{C}$ such that $\lim k_i = k$. We may assume that all of the k_i have the same orientation.

Let $f_i = (g \circ j) \circ k_i$, S a simple closed curve in the domain of f and $p \in T(S) \setminus f^{-1}(f(S))$. Note that $\lim f_i^{-1}(f_i(S)) \subset f^{-1}(f(S))$. Hence $p \in T(S) \setminus f_i^{-1}(f_i(S))$ for i sufficiently large. Moreover, since f_i converges to f, $f_i|_S$ is homotopic to $f|_S$ in the complement of $f(p)$ for i large. Thus for large i, $\text{degree}((f_i)_p) = \text{degree}(f_p)$, where

$$(f_i)_p(x) = \frac{f_i(x) - f_i(p)}{|f_i(x) - f_i(p)|} \text{ and } f_p(x) = \frac{f(x) - f(p)}{|f(x) - f(p)|}.$$

Since $g \circ j$ is an analytic map, it is positively oriented and we conclude that $\text{degree}((f_i)_p) = \text{degree}(f_p) > 0$ if k_i is orientation preserving and $\text{degree}((f_i)_p) = \text{degree}(f_p) < 0$ if k_i is orientation reversing. Thus, f is positively oriented if each k_i is orientation-preserving and f is negatively oriented if each k_i is orientation-reversing.

Suppose that X is a non-separating continuum and f is oriented. Let $y \in \partial f(X)$. Choose $y_i \in \partial f(X)$ and rays R_i, joining y_i to ∞ such that $R_i \cap f(X) = \{y_i\}$ and $\lim y_i = y$. Choose $x_i \in X$ such that $f(x_i) = y_i$. Since f is confluent, there exist closed and connected sets C_i, joining x_i to ∞ such that $C_i \cap X \subset f^{-1}(y_i)$. Hence there exist $x_i' \in f^{-1}(y_i) \cap \partial X$. We may assume that $\lim x_i' = x_\infty \in \partial X$ and $f(x_\infty) = y$ as desired. □

We shall need the following three results in the next section.

LEMMA 3.7.5. *Let X be a plane continuum and $f : \mathbb{C} \to \mathbb{C}$ a perfect, surjective map such that $f^{-1}(T(X)) = T(X)$ (i.e., $T(X)$ is fully invariant) and $f|_{\mathbb{C}\setminus T(X)}$ is confluent. Then for each $y \in \mathbb{C} \setminus T(X)$, each component of $f^{-1}(y)$ is acyclic.*

PROOF. Suppose there exists $y \in \mathbb{C} \setminus T(X)$ such that some component C of $f^{-1}(y)$ is not acyclic. Then there exists $z \in T(C) \setminus [f^{-1}(y) \cup T(X)]$. Then $T(X) \cup \{y\}$ does not separate $f(z)$ from infinity in $\mathbb{C}$. Let L be a ray in $\mathbb{C} \setminus [T(X) \cup \{y\}]$ from $f(z)$ to infinity. Then $L = \cup L_i$, where each $L_i \subset L$ is an arc with endpoint $f(z)$. For each i the component M_i of $f^{-1}(L_i)$ containing z maps onto L_i. Then $M = \cup M_i$ is a connected closed subset in $\mathbb{C} \setminus f^{-1}(y)$ from z to infinity. This is a contradiction since z is contained in a bounded complementary component of $f^{-1}(y)$. □

THEOREM 3.7.6. *Let X be a plane continuum and $f : \mathbb{C} \to \mathbb{C}$ a perfect, surjective map such that $f^{-1}(T(X)) = T(X)$ and $f|_{\mathbb{C}\setminus T(X)}$ is confluent. If A and B are crosscuts of $T(X)$ such that $B \cup X$ separates A from ∞ in $\mathbb{C}$, then $f(B) \cup T(X)$ separates $f(A) \setminus f(B)$ from ∞.*

PROOF. Suppose not. Then there exists a half-line L joining $f(A)$ to infinity in $\mathbb{C} \setminus (f(B) \cup T(X))$. As in the proof of Lemma 3.7.5, there exists a closed and connected set $M \subset \mathbb{C} \setminus (B \cup X)$ joining A to infinity, a contradiction. □

PROPOSITION 3.7.7. *Under the conditions of Theorem 3.7.6, if L is a ray irreducible from $T(X)$ to infinity, then each component of $f^{-1}(L)$ is closed in $\mathbb{C} \setminus X$ and is a connected set from X to infinity.*

3.8. Induced maps of prime ends

Suppose that $f : \mathbb{C} \to \mathbb{C}$ is an oriented perfect surjection and $f^{-1}(Y) = X$, where X and Y are acyclic continua and Y has no cutpoints. We will show that in this case the map f induces a confluent map F of the circle of prime ends of X to the circle of prime ends of Y. This result was announced by Mayer in the early 1980's but never appeared in print. It was also used (for homeomorphisms) by Cartwright and Littlewood in [**CL51**]. There are easy counterexamples that show if f is not confluent then it may not induce a continuous function between the circles of prime ends. For example, if $Y = \mathbb{D}$, X is the union of the unit disk and a copy of a half ray R which spirals to the unit circle and f is radial projection of R onto the unit circle, then f can be extended to a perfect map F of the plane so that $F^{-1}(Y) = X$ but F does not induce a continuous function from the circle of prime ends of X to the circle of prime ends of Y.

THEOREM 3.8.1. *Let X and Y be non-degenerate acyclic plane continua and $f : \mathbb{C} \to \mathbb{C}$ a perfect map such that:*
  (1) *Y has no cutpoint,*
  (2) *$f^{-1}(Y) = X$ and*
  (3) *$f|_{\mathbb{C}\setminus X}$ is confluent.*

3.8. INDUCED MAPS OF PRIME ENDS

Let $\varphi : \mathbb{D}^\infty \to \mathbb{C}^\infty \setminus X$ and $\psi : \mathbb{D}^\infty \to \mathbb{C}^\infty \setminus Y$ be conformal mappings. Define $\hat{f} : \mathbb{D}^\infty \to \mathbb{D}^\infty$ by $\hat{f} = \psi^{-1} \circ f \circ \varphi$.

Then $\hat{f}$ extends to a map $\bar{f} : \overline{\mathbb{D}^\infty} \to \overline{\mathbb{D}^\infty}$. Moreover, $\bar{f}^{-1}(\mathbb{S}^1) = \mathbb{S}^1$ and $F = \bar{f}|_{\mathbb{S}^1}$ is a confluent map.

PROOF. Note that f takes accessible points of X to accessible points of Y. For if P is a path in $[\mathbb{C} \setminus X] \cup \{p\}$ with endpoint $p \in X$, then by (2), $f(P)$ is a path in $[\mathbb{C} \setminus Y] \cup \{f(p)\}$ with endpoint $f(p) \in Y$.

Let A be a crosscut of X such that the diameter of $f(A)$ is less than half of the diameter of Y and let U be the bounded component of $\mathbb{C} \setminus (X \cup A)$. Let the endpoints of A be $x, y \in X$ and suppose that $f(x) = f(y)$. If x and y lie in the same component of $f^{-1}(f(x))$ then each crosscut $B \subset U$ of X is mapped to a generalized return cut of Y based at $f(x)$ (i.e., by (1) and (3) $f(\overline{U}) \cap Y = f(x)$ and the endpoints of B map to $f(x)$). Note that in this case by (1), $\partial f(U) \subset f(A) \cup \{f(x)\}$.

Now suppose that $f(x) = f(y)$ and x and y lie in distinct components of $f^{-1}(f(x))$. Then by unicoherence of $\mathbb{C}$, $\partial U \subset A \cup X$ is a connected set and $\partial U \not\subset \bar{A} \cup f^{-1}(f(x))$. Now $\partial U \setminus (\bar{A} \cup f^{-1}(f(x))) = \partial U \setminus f^{-1}(f(\bar{A}))$ is an open non-empty set in ∂U by (2). Thus there is a crosscut $B \subset U \setminus f^{-1}(f(\bar{A}))$ of X with $\bar{B} \setminus B \subset \partial U \setminus f^{-1}(f(\bar{A}))$. Now $f(B)$ is contained in a bounded component of $\mathbb{C} \setminus (Y \cup f(A)) = \mathbb{C} \setminus (Y \cup f(\bar{A}))$ by Theorem 3.7.6. Since $Y \cap f(\bar{A}) = \{f(x)\}$ is connected and Y does not separate $\mathbb{C}$, it follows by unicoherence that $f(B)$ lies in a bounded component of $\mathbb{C} \setminus f(\bar{A})$. Since $Y \setminus \{f(x)\}$ meets $f(\bar{B})$ and misses $f(\bar{A})$ and $Y \setminus \{f(x)\}$ is connected, $Y \setminus \{f(x)\}$ lies in a bounded complementary component of $f(\bar{A})$. This is impossible as the diameter of $f(A)$ is smaller than the diameter of Y. It follows that there exists a $\delta > 0$ such that if the diameter of A is less than δ and $f(x) = f(y)$, then x and y must lie in the same component of $f^{-1}(f(x))$.

In order to define the extension $\bar{f}$ of $\hat{f}$ over the boundary $\mathbb{S}^1$ of $\mathbb{D}^\infty$, let C_i be a chain of crosscuts of $\mathbb{D}^\infty$ which converge to a point $p \in \mathbb{S}^1$ such that $A_i = \varphi(C_i)$ is a null sequence of crosscuts or return cuts of X with endpoints a_i and b_i which converge to a point $x \in X$. There are three cases to consider:

Case 1. f identifies the endpoints of A_i for some A_i with diameter less than δ. In this case the chain of crosscuts is mapped by f to a sequence of generalized return cuts based at $f(a_i) = f(b_i) = f(x)$. Hence $f(a_i)$ is an accessible point of Y which corresponds (under ψ^{-1}) to a unique point $q \in \mathbb{S}^1$ (since Y has no cutpoints). Define $\bar{f}(p) = q$.

Case 2. Case 1 does not apply and there exists an infinite subsequence A_{i_j} of crosscuts such that $f(\bar{A}_{i_j}) \cap f(\bar{A}_{i_k}) = \emptyset$ for $j \neq k$. In this case $f(A_{i_j})$ is a chain of generalized crosscuts which converges to the point $f(x) \in Y$. The chain $\psi^{-1} \circ f(A_i)$ corresponds to a unique point $q \in \mathbb{S}^1$. Define $\bar{f}(p) = q$.

Case 3. Cases 1 and 2 do not apply. Without loss of generality suppose there exists an i such that for $j > i$ $f(\bar{A}_i) \cap f(\bar{A}_j)$ contains $f(a_i) = f(x)$. In this case $f(A_j)$ is a sequence of generalized crosscuts based at the accessible point $f(x)$ which corresponds to a unique point q on $\mathbb{S}^1$ as above. Define $\bar{f}(p) = q$.

It remains to be shown that $\bar{f}$ is a continuous extension of $\hat{f}$ and F is confluent. For continuity it suffices to show continuity at $\mathbb{S}^1$. Let $p \in \mathbb{S}^1$ and let C be a small crosscut of $\mathbb{D}^\infty$ whose endpoints are on opposite sides of p in $\mathbb{S}^1$ such that $A = \varphi(C)$ has diameter less than δ [Mil00] and such that the endpoints of A are two accessible points of X. Since f is uniformly continuous near X, the diameter of $f(A)$ is small and since ψ^{-1} is uniformly continuous with respect to connected sets in the complement of Y ([**UY51**]), the diameter of $B = \psi^{-1} \circ f \circ \varphi(C)$ is small. Also B is either a generalized crosscut or generalized return cut. Since $\hat{f}$ preserves separation of crosscuts, it follows that the image of the domain U bounded by C which does not contain ∞ is small. This implies continuity of $\bar{f}$ at p.

To see that F is confluent let $K \subset \mathbb{S}^1$ be a subcontinuum and let H be a component of $\bar{f}^{-1}(K)$. Choose a chain of crosscuts C_i such that $\varphi(C_i) = A_i$ is a crosscut of X meeting X in two accessible points a_i and b_i, $C_i \cap \bar{f}^{-1}(K) = \emptyset$ and $\lim C_i = H$. It follows from the preservation of crosscuts (see Theorem 3.7.6) that $\hat{f}(C_i)$ separates K from ∞. Hence $\hat{f}(C_i)$ must meet $\mathbb{S}^1$ on both sides of K and $\lim \bar{f}(C_i) = K$. Hence $F(H) = \lim \bar{f}(C_i) = K$ as required. □

COROLLARY 3.8.2. *Suppose that $f : \mathbb{C} \to \mathbb{C}$ is a perfect, oriented map of the plane, $X \subset \mathbb{C}$ is a subcontinuum without cut points and $f(X) = X$. Let $\hat{X}$ be the component of $f^{-1}(f(X))$ containing X. Let $\varphi : \mathbb{D}^\infty \to \mathbb{C}^\infty \setminus T(\hat{X})$ and $\psi : \mathbb{D}^\infty \to \mathbb{C}^\infty \setminus T(X)$ be conformal mappings. Define $\hat{f} : \mathbb{D}^\infty \setminus \varphi^{-1}(f^{-1}(X)) \to \mathbb{D}^\infty$ by $\hat{f} = \psi^{-1} \circ f \circ \varphi$. Put $\mathbb{S}^1 = \partial \mathbb{D}^\infty$.*

Then $\hat{f}$ extends over $\mathbb{S}^1$ to a map $\bar{f} : \overline{\mathbb{D}^\infty} \to \overline{\mathbb{D}^\infty}$. Moreover $\bar{f}^{-1}(\mathbb{S}^1) = \mathbb{S}^1$ and $F = \bar{f}|_{\mathbb{S}^1}$ is a confluent map.

PROOF. By Lemma 3.7.3 $f = g \circ m$ where m is a monotone perfect and onto mapping of the plane with acyclic point inverses, and g is an open and perfect surjection of the plane to itself. By Lemma 3.7.2, $f^{-1}(X)$ has finitely many components. It follows that there exist a simply connected open set V, containing $T(X)$, such that if U is the component of $f^{-1}(V)$ containing $\hat{X}$, then U contains no other components of $f^{-1}(X)$. It is easy to see that $f(U) = V$ and that U is simply connected. Hence U and V are homeomorphic to $\mathbb{C}$. Then $f|_U : U \to V$ is a confluent map. The result now follows from Theorem 3.8.1 applied to f restricted to U. □

CHAPTER 4

Partitions of domains in the sphere

4.1. Kulkarni-Pinkall Partitions

Throughout this section let K be a compact subset of the plane whose complement $U = \mathbb{C} \setminus K$ is connected. In the interest of completeness we define the Kulkarni-Pinkall partition of U and prove the basic properties of this partition that are essential for our work in Section 4.2. Kulkarni-Pinkall [**KP94**] worked in closed n-manifolds. We will follow their approach and adapt it to our situation in the plane.

We think of K as a closed subset of the Riemann sphere $\mathbb{C}^\infty$, with the spherical metric and set $U^\infty = \mathbb{C}^\infty \setminus K = U \cup \{\infty\}$. Let $\mathfrak{B}^\infty$ be the family of closed, round balls B in $\mathbb{C}^\infty$ such that $\text{Int}(B) \subset U^\infty$ and $|\partial B \cap K| \geq 2$. Then $\mathfrak{B}^\infty$ is in one-to-one correspondence with the family $\mathfrak{B}$ of closed subsets B of $\mathbb{C}$ which are the closure of a complementary component of a straight line or a round circle in $\mathbb{C}$ such that $\text{Int}(B) \subset U$ and $|\partial B \cap K| \geq 2$.

PROPOSITION 4.1.1. *If B_1 and B_2 are two closed round balls in $\mathbb{C}$ such that $B_1 \cap B_2 \neq \emptyset$ but does not contain a diameter of either B_1 or B_2, then $B_1 \cap B_2$ is contained in a ball of diameter strictly less than the diameters of both B_1 and B_2.*

PROOF. Let $\partial B_1 \cap \partial B_2 = \{s_1, s_2\}$. Then the closed ball with center $(s_1 + s_2)/2$ and radius $|s_1 - s_2|/2$ contains $B_1 \cap B_2$. □

If B is the closed ball of minimum diameter that contains K, then we say that B is the *smallest ball* containing K. It is unique by Proposition 4.1.1. It exists, since any sequence of balls of decreasing diameters that contain K has a convergent subsequence.

We denote the *Euclidean convex hull of K* by $\text{conv}_\mathcal{E}(K)$. It is the intersection of all closed half-planes (a closed half-plane is the closure of a component of the complement of a straight line) which contain K. Hence $p \in \text{conv}_\mathcal{E}(K)$ if p cannot be separated from K by a straight line.

Given a closed ball $B \in \mathfrak{B}^\infty$, $\text{int}(B)$ is conformally equivalent to the unit disk in $\mathbb{C}$. Hence its interior can be naturally equipped with the hyperbolic metric. Geodesics $\mathbf{g}$ in this metric are intersections of $\text{int}(B)$ with round circles $C \subset \mathbb{C}^\infty$ which perpendicularly cross the boundary ∂B. For every hyperbolic geodesic $\mathbf{g}$, $B \setminus \bar{\mathbf{g}}$ has exactly two components. We call the closure of such components *hyperbolic half-planes of B*. Given $B \in \mathfrak{B}^\infty$, the *hyperbolic convex hull of K in B* is the intersection of all (closed) hyperbolic half-planes of B which contain $K \cap B$ and we denote it by $\text{conv}_\mathcal{H}(B \cap K)$.

LEMMA 4.1.2. *Suppose that B is the smallest ball containing $K \subset \mathbb{C}$ and let $c \in B$ be its center. Then $c \in \text{conv}_\mathcal{H}(K \cap \partial B)$.*

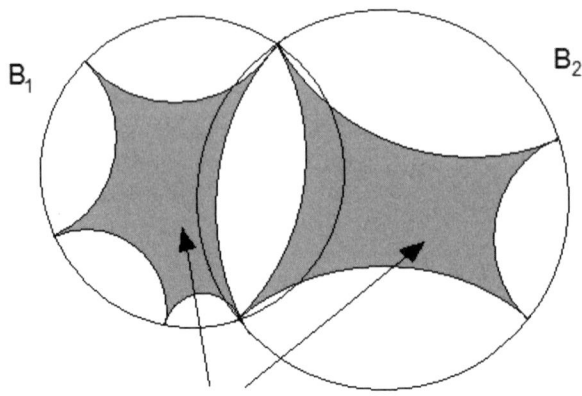

FIGURE 4.1. Maximal balls have disjoint hulls.

PROOF. By contradiction. Suppose that there exists a circle that separates the center c from $K \cap \partial B$ and crosses ∂B perpendicularly. Then there exists a line ℓ through c such that a half-plane bounded by ℓ contains $K \cap \partial B$ in its interior. Let $B' = B + v$ be a translation of B by a vector v that is orthogonal to ℓ and directed into this halfplane. If v is sufficiently small, then B' contains K in its interior. Hence, it can be shrunk to a strictly smaller ball that also contains K, contradicting that B has smallest diameter. □

LEMMA 4.1.3. Suppose that $B_1, B_2 \in \mathfrak{B}^\infty$ with $B_1 \neq B_2$. Then
$$\mathrm{conv}_\mathcal{H}(B_1 \cap \partial U) \cap \mathrm{conv}_\mathcal{H}(B_2 \cap \partial U) \subset \partial U.$$
In particular, $\mathrm{conv}_\mathcal{H}(B_1 \cap \partial U) \cap \mathrm{conv}_\mathcal{H}(B_2 \cap \partial U)$ contains at most two points.

PROOF. A picture easily explains this, see Figure 4.1. Note that $\partial U \cap [B_1 \cup B_2] \subset \partial(B_1 \cup B_1)$. Therefore $B_1 \cap \partial U$ and $B_2 \cap \partial U$ share at most two points. The open hyperbolic chords between these points in the respective balls are disjoint. □

It follows that any point in $U^\infty(X)$ can be contained in at most one hyperbolic convex hull. In the next lemma we see that each point of U^∞ is indeed contained in $\mathrm{conv}_\mathcal{H}(B \cap K)$ for some $B \in \mathfrak{B}^\infty$. So $\{U^\infty \cap \mathrm{conv}_\mathcal{H}(B \cap K) \mid B \in \mathfrak{B}^\infty\}$ is a partition of U^∞.

Since hyperbolic convex hulls are preserved by Möbius transformations, they are more easy to manipulate than the Euclidean convex hulls used by Bell (which are preserved only by Möbius transformations that fix ∞). This is illustrated by the proof of the following lemma.

LEMMA 4.1.4 (Kulkarni-Pinkall inversion lemma). For any $p \in \mathbb{C}^\infty \setminus K$ there exists $B \in \mathfrak{B}^\infty$ such that $p \in \mathrm{conv}_\mathcal{H}(B \cap K)$.

PROOF. We prove first that there exists $B^* \in \mathfrak{B}^\infty$ such that no circle which crosses ∂B^* perpendicularly separates $K \cap \partial B^*$ from ∞.

Let B' be the smallest round ball which contains K and let $B = \overline{\mathbb{C} \setminus B'}$. Then $B^* = B \cup \{\infty\} \in \mathfrak{B}^\infty$. If L is a circle which crosses $\partial B^* = \partial B'$ perpendicularly and separates $K \cap \partial B'$ from ∞, then it also separates $K \cap \partial B'$ from the center c' of B', contrary to Lemma 4.1.2. [To see this note that if F is the Möbius transformation which fixes points in the boundary of B' and interchanges the points ∞ and c', then $F(L) = L$. Hence it would follow that L separates c' from $K \cap \partial B'$, a contradiction with Lemma 4.1.2.] Hence, $\infty \in \text{conv}_\mathcal{H}(B^* \cap K)$.

Now let $p \in \mathbb{C}^\infty \setminus K$. Let $M : \mathbb{C}^\infty \to \mathbb{C}^\infty$ be a Möbius transformation such that $M(p) = \infty$. By the above argument there exists a ball $B^* \in \mathfrak{B}^\infty$ such that $\infty \in \text{conv}_\mathcal{H}(B^* \cap M(K))$. Then $B = M^{-1}(B^*) \in \mathfrak{B}^\infty$ and, since M preserves perpendicular circles, $p \in \text{conv}_\mathcal{H}(B \cap K)$ as desired. □

From Lemmas 4.1.3 and 4.1.4, we obtain the following Theorem which is a special case of a Theorem of Kulkarni and Pinkall [**KP94**].

THEOREM 4.1.5. *Suppose that $K \subset \mathbb{C}$ is a nondegenerate compact set such that its complement U^∞ in the Riemann sphere is connected. Then U^∞ is partitioned by the family*
$$\mathcal{KPP} = \{U^\infty \cap \text{conv}_\mathcal{H}(B \cap K) : B \in \mathfrak{B}^\infty\}.$$

Theorem 4.1.5 is the linchpin of the theory of geometric crosscuts. An analogue of it was known to Harold Bell and used by him implicitly since the early 1970's. Bell considered non-separating plane continua K and he used the equivalent notion of Euclidean convex hull of the sets $B \cap \partial U$ for all maximal balls $B \in \mathfrak{B}$ (see the comment following Theorem 4.2.5).

Let $B \in \mathfrak{B}^\infty$. If $B \cap \partial U^\infty(X)$ consists of two points a and b, then its (hyperbolic) hull is an open circular segment **g** with endpoints a and b and perpendicular to ∂B. We will call the crosscut **g** a $\mathcal{KP}$ *crosscut* or simply a $\mathcal{KP}$-*chord*. If $B \cap \partial U$ contains three or more points, then we say that the hull $\text{conv}_\mathcal{H}(B \cap \partial U)$ is a *gap*. A gap has nonempty interior. Its boundary in $\text{int}(B)$ is a union of open circular segments (with endpoints in K), which we also call $\mathcal{KP}$ *crosscuts or $\mathcal{KP}$-chords*. We denote by $\mathcal{KP}$ the collection of all open chords obtained as above using all $B \in \mathfrak{B}^\infty$.

The following example may serve to illustrate Theorem 4.1.5.
Example. Let K be the unit square $\{x + yi : -1 \le x, y \le 1\}$. There are five obvious members of $\mathfrak{B}$. These are the sets
$$\text{Im} z \ge 1, \ \text{Im} z \le -1, \ \text{Re} z \ge 1, \ \text{Re} z \le -1, \ |z| \ge \sqrt{2},$$
four of which are half-planes. These are the only members of $\mathfrak{B}$ whose hyperbolic convex hulls have non-empty interiors. However, for this example the family $\mathfrak{B}$ defined in the introduction of Section 4.1 is infinite. The hyperbolic hull of the half-plane $\text{Im} z \ge 1$ is the semi-disk $\{z \mid |z - i| \le 1, \ \text{Im} z > 1\}$. The hyperbolic hulls of the other three half-planes given above are also semi-disks. The hyperbolic hull of $|z| \ge \sqrt{2}$ is the unbounded region whose boundary consists of parts of four circles lying (except for their endpoints) outside K and contained in the circles of radius $\sqrt{2}$ and having centers at $-2, 2, -2i$ and $2i$, respectively. These hulls do not cover U as there are spaces between the hulls of the half-planes and the hull of $|z| \ge \sqrt{2}$.

If C is a circle that circumscribes K and contains exactly two of its vertices, such as $1 \pm i$, then the exterior ball B bounded by C is maximal. Now $\text{conv}_\mathcal{H}(B \cap K)$

is a single chord and the union of all such chords foliates the remaining spaces in $\mathbb{C} \setminus K$.

LEMMA 4.1.6. *If $\mathbf{g}_i$ is a sequence of $\mathcal{KP}$-chords with endpoints a_i and b_i, and $\lim a_i = a \neq b = \lim b_i$, then $\{\overline{\mathbf{g}_i}\}$ has a convergent subsequence and $\lim \overline{\mathbf{g}_{i_j}} = C$, where $\mathbf{g} = C \setminus \{a,b\} \in \mathcal{KP}$ is also a $\mathcal{KP}$-chord.*

PROOF. For each i let $B_i \in \mathfrak{B}^\infty$ such that $\mathbf{g}_i \subset \mathrm{conv}_\mathcal{H}(B_i \cap K)$. Then a subsequence B_{i_j} converges to some $B \in \mathfrak{B}^\infty$ and $\overline{\mathbf{g}_{i_j}}$ converges to a closed circular arc C in B with endpoints a and b, and C is perpendicular to ∂B. Hence $\mathbf{g} = C \setminus \{a,b\} \subset \mathrm{conv}_\mathcal{H}(B \cap K)$. So $\mathbf{g} \in \mathcal{KP}$. □

By Lemma 4.1.6, the family $\mathcal{KP}$ of chords has continuity properties similar to a foliation.

LEMMA 4.1.7. *For $a,b \in \partial U^\infty$, define $C(a,b)$ as the union of all $\mathcal{KP}$-chords with endpoints a and b. Then if $C(a,b) \neq \emptyset$, $C(a,b)$ is either a single chord, or $C(a,b) \cup \{a,b\}$ is a closed disk whose boundary consists of two $\mathcal{KP}$-chords contained in $C(a,b)$ together with $\{a,b\}$.*

PROOF. Suppose $\mathbf{g}$ and $\mathbf{h}$ are two distinct $\mathcal{KP}$-chords between a and b. Then $S = \mathbf{g} \cup \mathbf{h} \cup \{a,b\}$ is a simple closed curve. Choose a point z in the complementary domain V of S contained in U^∞. Since the hyperbolic hulls partition U^∞, there exists $B \in \mathfrak{B}^\infty$ such that $z \in \mathrm{conv}_\mathcal{H}(B \cap K)$. By Lemma 4.1.3, $\mathrm{conv}_\mathcal{H}(B \cap K)$ can only intersect $S \cap K$ in $\{a,b\}$. So $\mathrm{conv}_\mathcal{H}(B \cap K) \cap K = \{a,b\}$ and it follows that V is contained in $C(a,b)$.

The rest of the Lemma follows from 4.1.6. □

4.2. Hyperbolic foliation of simply connected domains

In this section we will apply the results from Section 4.1 to the case that K is a non-separating plane continuum (or, equivalently, that $U^\infty = \mathbb{C}^\infty \setminus K$ is simply connected). The results in this section are essential to [**OT07, OV09**] but are not used in this paper. The reader who is only interested in the fixed point question can skip this section.

Let $\mathbb{D}$ be the open unit disk in the plane. In this section we let $\phi : \mathbb{D} \to \mathbb{C}^\infty \setminus K = U^\infty$ be a Riemann map onto U^∞. We endow $\mathbb{D}$ with the hyperbolic metric, which is carried to U^∞ by the Riemann map. We use ϕ and the Kulkarni-Pinkall hulls to induce a closed collection Γ of chords in $\mathbb{D}$ that is a hyperbolic geodesic lamination in $\mathbb{D}$ (see [**Thu09**]).

Let $\mathbf{g} \in \mathcal{KP}$ be a chord with endpoints a and b. Then a and b are accessible points in K and $\overline{\phi^{-1}(\mathbf{g})}$ is an arc in $\mathbb{D}$ with endpoints $z,w \in \partial \mathbb{D}$. Let G be the hyperbolic geodesic in $\mathbb{D}$ joining z and w. Then G is an open circular arc which meets $\partial \mathbb{D}$ perpendicularly. Let Γ be the collection of all G such that $\mathbf{g} \in \mathcal{KP}$. We will prove that Γ inherits the properties of the family $\mathcal{KP}$ as described in Theorem 4.1.5 and Lemma 4.1.6 (see Lemma 4.2.3, Theorem 4.2.5 and the remark following 4.2.5).

Since members of $\mathcal{KP}$ do not intersect (though their closures are arcs which may have common endpoints) the same is true for distinct members of Γ. We will refer to the members of Γ (and their images under ϕ) as *hyperbolic chords* or *hyperbolic geodesics*. Given $\mathbf{g} \in \mathcal{KP}$ we denote the corresponding element of Γ by G and its image $\phi(\mathrm{G})$ in U^∞ by $\mathfrak{g}$. Note that Γ is a lamination of $\mathbb{D}$ in the sense of

Thurston[**Thu09**]. By a *gap* of Γ (or of $\phi(\Gamma)$), we mean the closure of a component of $\mathbb{D} \setminus \bigcup \Gamma$ in $\mathbb{D}$ (or its image under ϕ in U^∞, respectively).

LEMMA 4.2.1 (Jørgensen [**Pom92**, p.91 and 93r]). *Let B be a closed round ball such that its interior is in U^∞. Let $\gamma \subset \mathbb{D}$ be a hyperbolic geodesic. Then $\phi(\gamma) \cap B$ is connected. In particular, if R_t is an external ray in U^∞ and $B \in \mathfrak{B}^\infty$, then $R_t \cap B$ is connected.*

If $a, b \in \partial U^\infty$, recall that $C(a, b)$ is the union of all $\mathcal{KP}$-chords with endpoints a and b. From the viewpoint of prime ends, all chords in $C(a, b)$ are the same. That is why all the chords in $C(a, b)$ are replaced by a single hyperbolic chord $\mathfrak{g} \in \phi(\Gamma)$. The following lemma follows.

LEMMA 4.2.2. *Suppose $\mathbf{g} \in \mathcal{KP}$ and $\mathbf{g} \subset \mathrm{conv}_\mathcal{H}(B \cap \partial U^\infty)$ joins the points $a, b \in \partial U^\infty$ for some $B \in \mathfrak{B}^\infty$. If $G \in \Gamma$ is the corresponding hyperbolic geodesic, then $\mathfrak{g} = \phi(G) \subset B$.*

PROOF. We may assume that the Riemann map $\phi : \mathbb{D} \to U^\infty$ is extended over all points $x \in \mathbb{S}^1$ so that $\phi(x)$ is an accessible point of U^∞. Let $\phi^{-1}(a) = \hat{a}$, $\phi^{-1}(b) = \tilde{b}$ and $\phi^{-1}(B) = \tilde{B}$, and let G be the hyperbolic geodesic joining the points $\hat{a}$ and $\tilde{b}$ in $\mathbb{D}$. (Note that this extended map is not necessarily continuous at points of $\overline{\mathbb{D}}$ corresponding to accessible points of K.) Suppose, by way of contradiction, that $x \in G \setminus \tilde{B}$. Let C be the component of $\overline{\mathbb{D}} \setminus \overline{\phi^{-1}(\mathbf{g})}$ which does not contain x. Choose $a_i \to \hat{a}$ and $b_i \to \tilde{b}$ in $\mathbb{S}^1 \cap C$ and let H_i be the hyperbolic geodesic in $\mathbb{D}$ joining the points a_i and b_i. Then $\lim H_i = G$ and $H_i \cap \tilde{B}$ is not connected for i large. Hence $\phi(H_i) \cap B$ is not connected for i large. This contradiction with Lemma 4.2.1 completes the proof. □

LEMMA 4.2.3. *Suppose that $\{G_i\}$ is a sequence of hyperbolic chords in Γ and suppose that $x_i \in G_i$ such that $\{x_i\}$ converges to $x \in \mathbb{D}$. Then there is a unique hyperbolic chord $G \in \Gamma$ that contains x. Furthermore, $\lim G_i = \overline{G}$.*

PROOF. We may suppose that a subsequence sequence $\{G_{i_j}\}$ converges to a hyperbolic chord G which contains x. Let $\mathbf{g}_i \in \mathcal{KP}$ so that $\phi^{-1}(\mathbf{g}_i)$ is an open arc which joins the endpoints of G_i. By Lemma 4.1.6, there ia another subsequence so that $\lim \mathbf{g}_{i_{j(t)}} = \mathbf{g} \in \mathcal{KP}$. It follows that G is the hyperbolic chord joining the endpoints of $\phi^{-1}(\mathbf{g})$. Hence $G \in \Gamma$. Since the above argument applies to all subsequences, the sequence G_i must converge to G. □

So we have used the family of $\mathcal{KP}$-chords in U^∞ to stratify $\mathbb{D}$ to the family Γ of hyperbolic chords. In particular gaps of Γ are no longer necessarily disjoint but they can meet at most in a common boundary chord. By Lemma 4.2.2 for each $\mathcal{KP}$-chord $\mathbf{g} \subset \mathrm{conv}_\mathcal{H}(B \cap \partial U^\infty)$ its associated hyperbolic chord $\mathfrak{g} = \phi(G) \subset B$. Hence, there is a continuous deformation of U^∞ that maps $\bigcup \mathcal{KP}$ onto $\bigcup \phi(\Gamma)$, which suggests that components of $U^\infty \setminus \bigcup \phi(\Gamma)$ naturally correspond to the interiors of the gaps of the Kulkarni-Pinkall partition. That this is indeed the case is the substance of the next lemma.

LEMMA 4.2.4. *There is a $1-1$ correspondence between complementary domains $Z \subset \mathbb{D} \setminus \bigcup \Gamma$ and the interiors of Kulkarni-Pinkall gaps $\mathrm{conv}_\mathcal{H}(B \cap K)$. Moreover, for each gap Z of Γ there exists a unique $B \in \mathfrak{B}^\infty$ such that Z corresponds to the interior of the $\mathcal{KP}$ gap $\mathrm{conv}_\mathcal{H}(B \cap K) \cap U^\infty$ in that $\partial Z \cap \mathbb{D} = \bigcup \{G \in \Gamma \mid \mathbf{g} \in \mathcal{KP} \text{ and } \mathbf{g} \subset \partial \mathrm{conv}_\mathcal{H}(B \cap K)\}$ and $\phi(Z) \subset B$.*

PROOF. Let **g** and **h** be two distinct $\mathcal{KP}$-chords in the boundary of the gap $\text{conv}_{\mathcal{H}}(B \cap K)$ for some $B \in \mathfrak{B}^{\infty}$. Let $\{a, b\}$ and $\{c, d\}$ be the endpoints of $\phi^{-1}(\mathbf{g})$ and $\phi^{-1}(\mathbf{h})$, respectively. Since **g** and **h** are contained in the same gap, no hyperbolic chord of Γ separates G and H. Hence there exists a gap Z of Γ whose boundary includes the hyperbolic chords G and H. It now follows easily that for any $\mathbf{g}' \in \mathcal{KP}$ which is contained in the boundary of the same gap $\text{conv}_{\mathcal{H}}(B \cap K)$, G' is contained in the boundary of Z. Hence the $\mathcal{KP}$ gap $\text{conv}_{\mathcal{H}}(B \cap K)$ corresponds to the gap Z of Γ. Conversely, if Z is a gap of Γ in $\mathbb{D}$ then a similar argument, together with Lemmas 4.1.6 and 4.1.7, implies that Z corresponds to a unique gap $\text{conv}_{\mathcal{H}}(B \cap K)$ for some $B \in \mathfrak{B}^{\infty}$. The rest of the Lemma now follows from Lemma 4.2.2. □

So if $U^{\infty} = \mathbb{C}^{\infty} \setminus K$ is endowed with the hyperbolic metric induced by ϕ, then there exists a family of geodesic chords that share the same endpoints as elements of $\mathcal{KP}$. The complementary domains of $U^{\infty} \setminus \bigcup \{\mathfrak{g} \mid \mathfrak{g} \in \mathcal{KP}\}$ corresponds to the Kulkarni-Pinkall gaps. We summarize the results:

THEOREM 4.2.5. *Suppose that $K \subset \mathbb{C}$ is a non-separating continuum and let U^{∞} be its complementary domain in the Riemann sphere. There exists a family $\phi(\Gamma)$ of hyperbolic chords in the hyperbolic metric on U^{∞} such that for each $\mathfrak{g} \in \phi(\Gamma)$ there exists $B \in \mathfrak{B}^{\infty}$ and $\mathbf{g} \subset \text{conv}_{\mathcal{H}}(B \cap \partial U^{\infty})$ so that $\mathfrak{g}$ and $\mathbf{g}$ have the same endpoints and $\mathfrak{g} \subset B$. Each domain Z of $U^{\infty} \setminus \phi(\Gamma)$ naturally corresponds to a Kulkarni-Pinkall gap $\text{conv}_{\mathcal{H}}(B \cap \partial U^{\infty})$ The bounding hyperbolic chords of Z in U^{∞} correspond to the $\mathcal{KP}$-chords (i.e., chords in $\mathcal{KP}$) of $\text{conv}_{\mathcal{H}}(B \cap \partial U^{\infty})$.*

In order to obtain Bell's Euclidean foliation [**Bel76**] we could have modified the $\mathcal{KP}$ family as follows. Suppose that $B \in \mathfrak{B}$. Instead of replacing a $\mathcal{KP}$-chord $\mathbf{g} \in \text{conv}_{\mathcal{H}}(B \cap K)$ by a geodesic in the hyperbolic metric on U^{∞}, we could have replaced it by a straight line segment; i.e, the geodesic in the Euclidean metric. Then we would have obtained a family of open straight line segments. In so doing we would have replaced the gaps $\text{conv}_{\mathcal{H}}(B \cap \partial U^{\infty})$ by $\text{conv}_{\mathcal{E}}(B \cap \partial U^{\infty})$, which is the way in which Bell originally foliated $\text{conv}_{\mathcal{E}}(K) \setminus K$. We hope that the above argument provides a more transparent proof of Bell's result. Note that both in the hyperbolic and Euclidean case the elements of the foliation are not necessarily disjoint (hence we use the word "foliate" rather then " partition"). However, in both cases every point of U^{∞} is contained in either a unique chord or in the interior of a unique gap.

4.3. Schoenflies Theorem

In this short section we will show that the Schoenflies Theorem follows immediately from Theorem 4.2.5 (see [**Sie05**] for some recent history of this old problem and [**OT07, OV09**] for more details and extensions of these ideas). We want to emphasize here that no results of Chapter 3 are relied upon in Section 4.3.

THEOREM 4.3.1 (Schoenflies Theorem). *Suppose that $h : \mathbb{S}^1 \to \mathbb{C}$ is an embedding of the unit circle in the plane and U is a bounded complementary domain of $h(\mathbb{S}^1) = S$. Then there exists an embedding $H : \overline{\mathbb{D}} \to \mathbb{C}$ which extends h.*

PROOF. Let $\mathfrak{B} = \{B_{\alpha}\}$ be the collection of maximal open balls in U such that $|\partial B_{\alpha} \cap \partial U| \geq 2$. For each α let $F_{\alpha} = \text{conv}_{\mathcal{E}}(\partial B_{\alpha} \cap \partial U)$. Let $\mathcal{L}$ be the collection of all chords in the boundaries of all the sets F_{α} and let $\mathcal{L}^*$ be the union of all the chords in $\mathcal{L}$. Let $\mathcal{G} = \{G_\beta\}$ be the collection of all components of $U \setminus \mathcal{L}^*$. By Theorem 4.2.5 there exists for each $z \in U$ either a unique chord $\ell \in \mathcal{L}$ such that $z \in \ell$ or a unique $G_\beta \in \mathcal{G}$ such that $z \in G_\beta$. Moreover, in the latter case, there exists a unique α such that $\overline{G_\beta} = F_\alpha \subset \overline{B_\alpha}$. Note that all chords in $\mathcal{L}$ are straight line segments and all the sets F_α are Euclidean convex sets. Suppose that $x_i \in \mathcal{L}^*$ such that $\lim x_i = x_\infty$ and $x_i \in \ell_i \in \mathcal{L}$. Then either $\lim \ell_i = \ell_\infty$ and $x_\infty \in \ell_\infty \in \mathcal{L}$, or $\lim \ell_i = x_\infty \in \partial U \subset h(\mathbb{S}^1)$. Now we can pull back the lamination $\mathcal{L}$ to a lamination $\mathcal{E}$ of the unit disk $\mathbb{D}$ by h: if $\ell \in \mathcal{L}$ is a chord joining the points $y_1, y_2 \in S$, then connect the points $h^{-1}(y_1), h^{-1}(y_2)$ by the straight line segment, denoted by $h^{-1}(\ell)$ in the unit disk. Let $\mathcal{E}^*$ be the union of all such line segments. Note that each gap G_β of U uniquely corresponds to a (Euclidean convex) gap H_β of $\mathcal{E}$ (i.e., H_β is a component of $\mathbb{D} \setminus \mathcal{E}^*$).

Now extend h first over $\mathcal{E}^*$ by mapping each chord in $\mathcal{E}$ linearly onto the corresponding chord in $\mathcal{L}$. For each gap H_β (G_β) of $\mathcal{E}$ ($\mathcal{L}$) let h_β (g_β, respectively) be its barycenter. Then it follows easily that we can extend the map h continuously by defining $H(h_\beta) = g_\beta$ for each β. Finally extend H over all of H_β by mapping, for each $w \in \partial H_\beta$ the straight line segment wh_β linearly onto the straight line segment joining the points $H(w)$ and $H(h_\beta) = g_\beta$. Then H is the required extension of h. □

4.4. Prime ends

We will follow the notation from Section 4.1 in the case that $K = T(X)$ where X is a plane continuum. Here we assume, as in the introduction to this paper, that $f : \mathbb{C} \to \mathbb{C}$ takes the continuum X into $T(X)$ with no fixed points in $T(X)$, and X is minimal with respect to these properties. We apply the Kulkarni-Pinkall partition to $U^\infty = \mathbb{C}^\infty \setminus T(X)$. Recall that $\mathcal{KPP} = \{\text{conv}_\mathcal{H}(B \cap K) \cap U^\infty \mid B \in \mathfrak{B}^\infty\}$ is the Kulkarni Pinkall partition of U^∞ as given by Theorem 4.1.5.

Let $B^\infty \in \mathfrak{B}^\infty$ be the maximal ball such that $\infty \in \text{conv}_\mathcal{H}(B^\infty \cap K)$. As before we use balls on the sphere. In particular, straight lines in the plane correspond to circles on the sphere containing the point at infinity. The subfamily of $\mathcal{KPP}$ whose elements are of diameter $\le \delta$ in the spherical metric is denoted by $\mathcal{KPP}_\delta$. The subfamily of chords in $\mathcal{KP}$ of diameter $\le \delta$ is denoted by $\mathcal{KP}_\delta$.

By Lemma 4.1.6 we know that the families $\mathcal{KP}$ and $\mathcal{KPP}$ have nice continuity properties. However, $\mathcal{KP}$ and $\mathcal{KPP}$ are not closed in the hyperspace of compact subsets of $\mathbb{C}^\infty$: a sequence of chords or hulls may converge to a point in the boundary of U^∞ (in which case it must be a null sequence).

PROPOSITION 4.4.1 (Closedness). Let $\{\mathbf{g}_i\}$ be a convergent sequence of distinct elements in $\mathcal{KP}_\delta$, then either $\mathbf{g}_i$ converges to a chord $\mathbf{g}$ in $\mathcal{KP}_\delta$ or $\mathbf{g}_i$ converges to a point of X. In the first case, for large i and δ sufficiently small, $\text{var}(f, \mathbf{g}, T(X)) = \text{var}(f, \mathbf{g}_i, T(X))$.

PROOF. By Lemma 4.1.6, we know that the first conclusion holds if $\mathbf{g} = \lim \mathbf{g}_i$ contains a point of U^∞. Hence we only need to consider the case when $\lim \mathbf{g}_i = \mathbf{g} \subset \partial U^\infty \subset T(X)$. If the diameter of $\mathbf{g}_i$ converged to zero, then $\mathbf{g}$ is a point as desired. Assume that this is not the case and let B_i be the maximal ball that contains $\mathbf{g}_i$. Under our assumption, the diameters of $\{B_i\}$ do not decay to zero. Let $B \in \mathfrak{B}^\infty$ be the limit of a subsequence B_{i_j}. Then $\lim \mathbf{g}_i$ is a piece of a round circle which crosses ∂B perpendicularly. Hence $\lim \mathbf{g}_i \cap \text{int}(B) \ne \emptyset$, contradicting the fact that $\mathbf{g} \subset \partial U^\infty \subset T(X)$. Note that for δ sufficiently small, $\overline{\mathbf{g}} \cap f(\overline{\mathbf{g}}) = \emptyset$. Hence, var$(f, \mathbf{g}, T(X))$ and var$(f, \mathbf{g}_i, T(X))$ are defined for all i sufficiently large. Then last statement in the Lemma follows from stability of variation (see Section 2.2). □

COROLLARY 4.4.2. *For each $\varepsilon > 0$, there exist $\delta > 0$ such that for all $\mathbf{g} \in \mathcal{KP}$ with $\mathbf{g} \subset B(T(X), \delta)$, diam$(\mathbf{g}) < \varepsilon$.*

PROOF. Suppose not, then there exist $\varepsilon > 0$ and a sequence $\mathbf{g}_i$ in $\mathcal{KP}$ such that $\lim \mathbf{g}_i \subset X$ and diam$(\mathbf{g}_i) \geq \varepsilon$ a contradiction to Proposition 4.4.1. □

The proof of the following well-known proposition is omitted.

PROPOSITION 4.4.3. *For each $\varepsilon > 0$ there exists $\delta > 0$ such that for each open arc A with distinct endpoints a, b such that $\overline{A} \cap T(X) = \{a, b\}$ and diam$(A) < \delta$, $T(T(X) \cup A) \subset B(T(X), \varepsilon)$.*

PROPOSITION 4.4.4. *Let ε, δ be as in Proposition 4.4.3 above with $\delta < \varepsilon/2$ and let $B \in \mathfrak{B}^\infty$. Let A be a crosscut of $T(X)$ such that diam$(A) < \delta$. If $x \in T(A \cup T(X)) \cap \operatorname{conv}_\mathcal{H}(B \cap T(X)) \setminus T(X)$ and $d(x, A) \geq \varepsilon$, then the radius of B is less than ε. Hence, diam$(\operatorname{conv}_\mathcal{H}(B \cap T(X))) < 2\varepsilon$.*

PROOF. Let z be the center of B. If $d(z, T(X)) < \varepsilon$ then diam$(B) < 2\varepsilon$ and we are done. Hence, we may assume that $d(z, T(X)) \geq \varepsilon$. We will show that this leads to a contradiction. By Proposition 4.4.3 and our choice of δ, $z \in \mathbb{C}^\infty \setminus T(A \cup X)$. The straight line segment ℓ from x to z must cross $T(X) \cup A$ at some point w. Since the segment ℓ is in the interior of the maximal ball B, it is disjoint from $T(X)$, so $w \in A$. Hence $d(x, w) \geq \varepsilon$ and, since $x \in B$, $B(w, \varepsilon) \subset B$. This is a contradiction since $A \subset B(w, \delta)$ and $\delta < \varepsilon/2$ so $\overline{A}$ would be contained in the interior of B which is impossible since A is a crosscut of $T(X)$. □

PROPOSITION 4.4.5. *Let C be a crosscut of $T(X)$ and let A and B be disjoint closed sets in $T(X)$ such that $\overline{C} \cap A \neq \emptyset \neq \overline{C} \cap B$. For each $x \in C$, let $F_x \in \mathcal{KPP}$ so that $x \in F_x$. If each $\overline{F_x}$ intersects $A \cup B$, then there exists an $F_\infty \in \mathcal{KPP}$ such that $\overline{F_\infty}$ intersects A, B and $\overline{C}$.*

PROOF. Let $a \in A, b \in B$ be the endpoints of $\overline{C}$. Let $C_a, C_b \subset C$ be the set of points $x \in C$ such that $\overline{F_x}$ intersects A or B, respectively. Then C_a and C_b are closed subsets by Proposition 4.4.1. Note that $d(A, B) > 0$. If $C_a = \emptyset$, choose $x_i \in C$ converging to $a \in A \cap \overline{C}$. Let $F_{x_i} = \operatorname{conv}_\mathcal{H}(B_i \cap T(X))$, where $B_i \in \mathfrak{B}$ and assume that $B_\infty = \lim B_i$. Then by Lemma 4.1.6, $\overline{F_\infty} \cap B \neq \emptyset$ and $\lim F_{x_i} \subset \overline{F_\infty} \subset \operatorname{conv}_\mathcal{H}(B_\infty \cap K)$. Then $\overline{F_\infty} \cap B \neq \emptyset$ and $a \in A \cap \overline{C} \cap \overline{F_\infty}$. Suppose now $C_a \neq \emptyset \neq C_b$. Then C_a and C_b are closed and, since C is connected, $C_a \cap C_b \neq \emptyset$. Let $y \in C_a \cap C_b$. Then $\overline{F_y} \cap A \neq \emptyset \neq \overline{F_y} \cap B$ and $y \in F_y \cap C$. □

Proposition 4.4.5 allows us to replace small crosscuts which essentially cross the external ray R_t with non-trivial principal continuum with small nearby $\mathcal{KP}$-chords which also essentially crosses R_t. For if C is a small crosscut of $T(X)$ with endpoints a and b which crosses the external ray R_t essentially, let A and B be the closures of the sets in $T(X)$ accessible from a and b, respectively by small arcs missing R_t. If the F_∞ of proposition 4.4.5 is a gap $\operatorname{conv}_\mathcal{H}(B \cap T(X))$, then a $\mathcal{KP}$-chord in its boundary crosses R_t essentially.

Fix a Riemann map $\varphi : \mathbb{D}^\infty \to U^\infty = \mathbb{C}^\infty \setminus T(X)$ with $\varphi(\infty) = \infty$. Recall that an external ray R_t is the image of the radial line segment with argument $2\pi t i$ under the map φ.

4.4. PRIME ENDS

PROPOSITION 4.4.6. Suppose the external ray R_t lands on $x \in T(X)$, and $\{\mathbf{g}_i\}_{i=1}^\infty$ is a sequence of crosscuts of $T(X)$ converging to x such that there exists a null sequence of arcs $A_i \subset \mathbb{C} \setminus T(X)$ joining $\mathbf{g}_i$ to R_t. Then for sufficiently large i, $\operatorname{var}(f, \mathbf{g}_i, T(X)) = 0$.

PROOF. Since f is fixed point free on $T(X)$ and $f(x) \in T(X)$, we may choose a small ball W with center x in $\mathbb{C}$ such that $f(\overline{W}) \cap (\overline{W} \cup R_t) = \emptyset$. For sufficiently large i, $A_i \cup \mathbf{g}_i \subset W$. Then for each such i there exists a junction J_i starting from a point in $\mathbf{g}_i$, with all of its legs staying in W close to A_i until it reaches R_t, and then staying close to R_t to ∞. By our choice of W, $\operatorname{var}(f, \mathbf{g}_i, T(X)) = 0$. □

PROPOSITION 4.4.7. Suppose that for an external ray R_t we have $R_t \cap \operatorname{int}(\operatorname{conv}_\mathcal{E}(T(X))) \neq \emptyset$. Then there exists $x \in R_t$ such that the $(T(X), x)$-end of R_t is contained in $\operatorname{conv}_\mathcal{E}(T(X))$. In particular there exists a chord $\mathbf{g} \in \mathcal{KP}$ such that R_t crosses $\mathbf{g}$ essentially.

PROOF. External rays in U^∞ correspond to geodesic half-lines starting at infinity in the hyperbolic metric on $\mathbb{C}^\infty \setminus T(X)$. Half-planes are conformally equivalent to disks. Therefore, Jørgensen's lemma applies: the intersection of R_t with a half-plane is connected, so it is a half-line. Since the Euclidean convex hull of $T(X)$ is the intersection of all half-planes containing $T(X)$, $R_t \cap \operatorname{conv}_\mathcal{E}(T(X))$ is connected. □

LEMMA 4.4.8. Let $\mathcal{E}_t$ be a channel (that is, a prime end such that $\operatorname{Pr}(\mathcal{E}_t)$ is non-degenerate) in $T(X)$. Then for each $x \in \operatorname{Pr}(\mathcal{E}_t)$, for every $\delta > 0$, there is a chain $\{\mathbf{g}_i\}_{i=1}^\infty$ of chords defining $\mathcal{E}_t$ selected from $\mathcal{KP}_\delta$ with $\mathbf{g}_i \to x \in \partial T(X)$.

PROOF. Let $x \in \operatorname{Pr}(\mathcal{E}_t)$ and let $\{C_i\}$ be a defining chain of crosscuts for $\operatorname{Pr}(\mathcal{E}_t)$ with $\{x\} = \lim C_i$. By Proposition 4.4.5, in particular by the remark following the proof of that proposition, there is a sequence $\{\mathbf{g}_i\}$ of $\mathcal{KP}$-chords such that $d(\mathbf{g}_i, C_i) \to 0$ and R_t crosses each $\mathbf{g}_i$ essentially. By Proposition 4.4.4, the sequence $\mathbf{g}_i$ converges to $\{x\}$. □

LEMMA 4.4.9. Suppose an external ray R_t lands on $a \in T(X)$ with $\{a\} = \operatorname{Pr}(\mathcal{E}_t) \neq \operatorname{Im}(\mathcal{E}_t)$. Suppose $\{x_i\}_{i=1}^\infty$ is a collection of points in U^∞ with $x_i \to x \in \operatorname{Im}(\mathcal{E}_t) \setminus \{a\}$ and $\phi^{-1}(x_i) \to t$. Then there is a sequence of $\mathcal{KP}$-chords $\{\mathbf{g}_i\}_{i=1}^\infty$ such that for sufficiently large i, $\mathbf{g}_i$ separates x_i from ∞, $\mathbf{g}_i \to a$ and $\phi^{-1}(\mathbf{g}_i) \to t$.

PROOF. The existence of the chords $\mathbf{g}_i$ again follows from the remark following Proposition 4.4.5. It is easy to see that $\lim \varphi^{-1}(\mathbf{g}_i) \to t$. □

4.4.1. Auxiliary Continua.
We use $\mathcal{KP}$-chords to form Carathéodory loops around the continuum $T(X)$.

DEFINITION 4.4.10. Fix $\delta > 0$. Define the following collections of chords:

$$\mathcal{KP}_\delta^+ = \{\mathbf{g} \in \mathcal{KP}_\delta \mid \operatorname{var}(f, \mathbf{g}, T(X)) \geq 0\}$$
$$\mathcal{KP}_\delta^- = \{\mathbf{g} \in \mathcal{KP}_\delta \mid \operatorname{var}(f, \mathbf{g}, T(X)) \leq 0\}$$
$$\mathcal{KP}_\delta = \mathcal{KP}_\delta^+ \cup \mathcal{KP}_\delta^-.$$

To each collection of chords above, there corresponds an auxiliary continuum defined as follows:

$$T(X)_\delta = T(X \cup (\cup \mathcal{KP}_\delta))$$

$$T(X)_\delta^+ = T(X \cup (\cup \mathcal{KP}_\delta^+))$$
$$T(X)_\delta^- = T(X \cup (\cup \mathcal{KP}_\delta^-))$$

PROPOSITION 4.4.11. *Let $Z \in \{T(X)_\delta, T(X)_\delta^+, T(X)_\delta^-\}$, and correspondingly $\mathcal{W} \in \{\mathcal{KP}_\delta, \mathcal{KP}_\delta^+, \mathcal{KP}_\delta^-\}$. Then the following hold:*

(1) *Z is a non-separating plane continuum.*
(2) *$\partial Z \subset T(X) \cup (\cup \mathcal{W})$.*
(3) *Every accessible point y in ∂Z is either a point of $T(X)$ or a point interior to a chord $\mathbf{g} \in \mathcal{W}$.*
(4) *If $y \in \partial Z \cap \mathbf{g}$ with $\mathbf{g} \in \mathcal{W}$, then y is accessible, $\mathbf{g} \subset \partial Z$ and ∂Z is locally connected at each point of $\mathbf{g}$. Hence, if $\varphi : \mathbb{D}^\infty \to \mathbb{C}^\infty \setminus Z$ is the Riemann map and R_t is an external ray landing at y, then φ extends continuously to an open interval in $\mathbb{S}^1$ containing t. Moreover, if $y \in \partial Z \cap [\overline{\mathbf{g}} \setminus \mathbf{g}]$, then φ extends continuously over a half open $J \subset \mathbb{S}^1$ with endpoint t so that $\varphi(J) \subset \overline{\mathbf{g}}$.*

PROOF. By Proposition 4.4.1, $T(X) \cup (\cup \mathcal{W})$ is compact. Moreover, $T(X) \cup (\cup \mathcal{W})$ is connected since each crosscut $A \in \mathcal{W}$ has endpoints in $T(X)$. Hence, the topological hull $T(T(X) \cup (\cup \mathcal{W}))$ is a non-separating plane continuum, establishing (1).

Since Z is the topological hull of $T(X) \cup (\cup \mathcal{W})$, no boundary points can be in complementary domains of $T(X) \cup (\cup \mathcal{W})$. Hence, $\partial Z \subset T(X) \cup (\cup \mathcal{W})$, establishing (2). Conclusion (3) follows immediately.

Suppose $y \in \partial Z \cap \mathbf{g}$ with $\mathbf{g} \in \mathcal{W}$. Then $\text{Sh}(\mathbf{g}) \subset Z$ and there exists $y_i \in \mathbb{C} \setminus Z$ such that $\lim y_i = y$. We may assume that all the points y_i are on the "same side" of the arc $\mathbf{g}$ (i.e., $y_i \in \mathbb{C} \setminus \text{Sh}(\mathbf{g})$). This side of $\mathbf{g}$ is either (1) a limit of $\mathcal{KP}$-chords $\mathbf{g}_j$, or (2) there exists a gap $\text{conv}_\mathcal{H}(B \cap X)$ on this side with $\mathbf{g}$ in its boundary. In case (1), $\mathbf{g} \subset \text{Sh}(\mathbf{g}_j)$ and, since $y_i \in \mathbb{C} \setminus Z$ for all i, $\mathbf{g}_j \notin \mathcal{W}$. Hence each $\mathbf{g}_j \subset \mathbb{C} \setminus Z$ for all j. It follows that every point of $\mathbf{g}$ is accessible, $\mathbf{g} \subset \partial Z$ and ∂Z is locally connected at each point of $\mathbf{g}$. In case (2) there exists a chord $\mathbf{g}' \neq \mathbf{g}$ in the boundary of $\text{conv}_\mathcal{H}(B \cap X)$ which separates $\mathbf{g}$ from infinity. Then $\mathbf{g}' \notin \mathcal{W}$ and the interior of $\text{conv}_\mathcal{H}(B \cap X) \subset \mathbb{C} \setminus Z$. Hence the same conclusion follows.

The last part of (4) follows from the proof of Carathéodory's theorem (see [**Pom92**]). □

PROPOSITION 4.4.12. *$T(X)_\delta$ is locally connected; hence, $\partial T(X)_\delta$ is a Carathéodory loop.*

PROOF. Suppose that $T(X)_\delta$ is not locally connected. Then $T(X)_\delta$ has a non-trivial impression and there exist $0 < \varepsilon < \delta/2$ and a chain A_i of crosscuts of $T(X)_\delta$ such that $\text{diam}(\text{Sh}(A_i)) > \varepsilon$ for all i. We may assume that $\lim A_i = y \in T(X)_\delta$.

By Proposition 4.4.11 (4) we may assume $y \in X$. Choose $z_i \in \text{Sh}(A_i)$ such that $d(z_i, y) > \varepsilon$. We can enlarge the crosscut A_i of $T(X)_\delta$ to a crosscut C_i of $T(X)$ as follows. Suppose that A_i joins the points a_i^+ and a_i^- in $T(X)_\delta$. If $a_i^+ \in T(X)$, put $y_i^+ = a_i^+$. Otherwise a_i^+ is contained in a chord $\mathbf{g}_i^+ \in \mathcal{KP}_\delta$, with endpoints in $T(X)$, which is contained in $T(X)_\delta$. Since $\lim A_i = y$, we can select one of these endpoints and call it y_i^+ such that $d(y_i^+, a_i^+) \to 0$. Define $\mathbf{g}_i^-$ and y_i^- similarly. Then $\mathbf{g}_i^+ \cup A_i \cup \mathbf{g}_i^-$ contains a crosscut C_i of $T(X)$ joining the points y_i^+ and y_i^- such that

$\lim C_i = y$. We claim that $z_i \in \text{Sh}(C_i)$. To see this note that, since $z_i \in \text{Sh}(A_i)$, there exists a half-ray $R_i \subset \mathbb{C} \setminus T(X)_\delta$ joining z_i to infinity such that $|R_i \cap A_i|$ is an odd number and each intersection is transverse. Since $R_i \cap C_i = R_i \cap A_i$ it follows that $z_i \in \text{Sh}(C_i)$. Let $\text{conv}_\mathcal{H}(B_i \cap X)$ be the unique hull of the Kulkarni-Pinkall partition $\mathcal{KPP}$ which contains z_i. Since $\text{diam}(C_i) \to 0$ and $d(z_i, y) > \varepsilon$, it follows from Proposition 4.4.4 that $\text{diam}(\text{conv}_\mathcal{H}(B_i \cap X)) < 2\varepsilon < \delta$. This contradicts the fact that $z_i \in \mathbb{C} \setminus T(X)_\delta$ and completes the proof. □

Part 2

Applications of Basic Theory

CHAPTER 5

Description of main results of Part 2

We begin by describing the results obtained in Part 2. These results are applications of the tools developed in Part 1. We will say that a continuum X is *decomposable* if there exist two proper subcontinua A, B of X such that $X = A \cup B$. A continuum which is not decomposable, is called *indecomposable*.

5.1. Outchannels

In Chapter 6 we will study outchannels. Outchannels were introduced by Bell to establish that a minimal counterexample to the Plane Fixed Point Problem must be an indecomposable continuum. In Chapter 6 we will recover this result and strengthen it by showing that the outchannel in a minimal counterexample to the Plane Fixed Point Problem is unique: there exists exactly one prime end $\mathcal{E}_t$ which corresponds to a dense channel with non zero variation. It will follow that the variation of this channel must be -1 while all other small crosscuts, which do not cross this channel essentially, must have variation zero. Let us assume that $f : \mathbb{C} \to \mathbb{C}$ with a forward invariant non-separating continuum X presents a (possibly existing) minimal counterexample to the Plane Fixed Point Problem.

We construct a specific locally connected (but not invariant) continuum $X' \supset X$ by adding small crosscuts to X. This will be done in a careful way; we will only add Kulkarni-Pinkall crosscuts from $\mathcal{KP}$. This construction is used to show that if there is a minimal counterexample (X, f) to the Plane Fixed Point Problem, then there exists a continuum Z such that the following facts hold.

(1) $Z \supset X$;
(2) there exists a one-to-one map $\varphi : \mathbb{R} \to Z$,
(3) $\varphi(\mathbb{R})$ is the set of accessible points of Z,
(4) as $t \to \infty$, $\varphi(t)$ and $\varphi(-t)$ run along opposite sides of the outchannel.

Moreover, the same construction is important in the proof of the uniqueness of the outchannel.

These ideas are also applied in [**BO09**]. There it was shown that in certain cases a minimal subcontinuum X without a fixed point must be fully invariant. As an important tool it was shown in that paper that the map f can be modified on $\mathbb{C} \setminus X$ to a map g such that $g(R_t) = R_t$, g maps points on R_t closer to infinity and g locally interchanges the two sides of R_t. Here R_t is the conformal external ray which represents the prime end corresponding to the outchannel. Note that if X is fully invariant then a prime end which corresponds to the outchannel has the property that, in a defining sequence $\{C_i\}$ of crosscuts of the prime end $f(C_i)$ separates C_i from infinity in $U^\infty(X)$ (thus justifying the name "outchannel").

Suppose that $f : \mathbb{C} \to \mathbb{C}$ is a perfect map, X is a continuum, f has no fixed point in $T(X)$ and X is minimal with respect to $f(X) \subset T(X)$. Fix $\eta > 0$ such that for each $\mathcal{KP}$-chord $\mathbf{g} \subset T(X)_\eta$, $\overline{\mathbf{g}} \cap f(\overline{\mathbf{g}}) = \emptyset$ and f is fixed point free on $T(X)_\eta$. In this case we will say that η *defines variation near* X and that the triple (f, X, η) satisfies the *standing hypothesis*. As usual, for a continuum X let $U^\infty = \mathbb{C}^\infty \setminus T(X)$.

DEFINITION 5.1.1 (Outchannel). Suppose that the triple (X, f, η) satisfies the standing hypothesis. An *outchannel* of the non-separating plane continuum $T(X)$ is a prime end $\mathcal{E}_t$ of U^∞ such that for some chain $\{\mathbf{g}_i\}$ of crosscuts defining $\mathcal{E}_t$, $\mathrm{var}(f, \mathbf{g}_i, T(X)) \neq 0$ for every i. We call an outchannel $\mathcal{E}_t$ of $T(X)$ a *geometric outchannel* if and only if for sufficiently small δ, every chord in $\mathcal{KP}_\delta$, which crosses $\mathcal{E}_t$ essentially, has nonzero variation. We call a geometric outchannel *negative* (respectively, *positive*) (starting at $\mathbf{g} \in \mathcal{KP}$) if and only if every $\mathcal{KP}$-chord $\mathbf{h} \subset T(X)_\eta \cap \overline{\mathrm{Sh}(\mathbf{g})}$, which crosses $\mathcal{E}_t$ essentially, has negative (respectively, positive) variation.

5.2. Fixed points in invariant continua

In this Section we describe the results obtained in Section 7.1 of Chapter 7. The main result of Section 7.1 solves the Plane Fixed Point Problem in the affirmative for positively oriented maps of the plane. Namely, the following theorem is proven.

THEOREM 7.1.3. *Suppose $f : \mathbb{C} \to \mathbb{C}$ is a positively oriented map and X is a continuum such that $f(X) \subset T(X)$. Then there exists a point $x_0 \in T(X)$ such that $f(x_0) = x_0$.*

5.3. Fixed points in non-invariant continua – the case of dendrites

As described in Chapter 1, in the rest of Chapter 7 we want to extend Theorem 7.1.3 to at least some non-invariant continua. We are motivated by the interval case in which to conclude that there exists a fixed point in an interval it is enough to know that the endpoints of the interval map in opposite directions, and the invariantness of the interval itself is not crucial.

In Section 7.2 we extend, in the spirit of the interval case, a well-known result according to which a map of a dendrite into itself has a fixed point (Theorem 1.0.2, see [**Nad92**]). We show the existence of fixed points in non-invariant dendrites and, with some additional conditions, obtain also results related to the number of periodic points of f. To state the precise results we need some definitions.

DEFINITION 5.3.1 (Boundary scrambling for dendrites). Suppose that f maps a dendrite D_1 to a dendrite $D_2 \supset D_1$. Put $E = \overline{D_2 \setminus D_1} \cap D_1$ (observe that E may be infinite). If for each *non-fixed* point $e \in E$, $f(e)$ is contained in a component of $D_2 \setminus \{e\}$ which intersects D_1, then we say that f has the *boundary scrambling property* or that it *scrambles the boundary*. Observe that if D_1 is invariant then f automatically scrambles the boundary.

The following theorem is the first result obtained in Section 7.2.

THEOREM 7.2.2. *Let $f : D_1 \to D_2$ be a map, where D_1 and D_2 are dendrites and $D_1 \subset D_2$. The following claims hold.*

(1) *If $a, b \in D_1$ are such that a separates $f(a)$ from b and b separates $f(b)$ from a, then there exists a fixed point $c \in (a, b)$. Thus, if $e_1 \neq e_2 \in E$ are such that each $f(e_i)$ belongs to a component of $D_2 \setminus \{e_i\}$ disjoint from D_1 then there is a fixed point $c \in (e_1, e_2)$.*

(2) If f scrambles the boundary, then f has a fixed point.

To give the next definition we recall that if $x \in Y$ then the *valence of Y at x*, $\mathrm{val}_Y(x)$, is defined as the number of connected components of $Y \setminus \{x\}$, and x is said to be a *cutpoint (of Y)* if $\mathrm{val}_Y(x) > 1$.

DEFINITION 5.3.2 (Weakly repelling periodic points). In the situation of Definition 5.3.1, let $a \in D_1$ be a fixed point and suppose that there exists a component B of $D_1 \setminus \{a\}$ such that arbitrarily close to a in B there exist fixed cutpoints of D_1 or points x separating a from $f(x)$. Then we say that a is a *weakly repelling fixed point (of f in B)*. A periodic point $a \in D_1$ is said to be simply *weakly repelling* if there exists n and a component B of $D_1 \setminus \{a\}$ such that a is a weakly repelling fixed point of f^n in B.

We use the notions introduced in Definition 5.3.2 to prove Theorem 7.2.6.

THEOREM 7.2.6. Suppose that $f : D \to D$ is continuous where D is a dendrite and all its periodic points are weakly repelling. Then f has infinitely many periodic cutpoints.

This theorem is applied in Theorem 7.2.7 where it is shown that if $g : J \to J$ is a *topological polynomial* on its dendritic Julia set (e.g., if f is a complex polynomial with a dendritic Julia set) then it has infinitely many periodic cutpoints.

5.4. Fixed points in non-invariant continua – the planar case

In parallel with the dendrite case, we want to extend Theorem 7.1.3 to a larger class of maps of the plane and non-invariant continua such that certain "boundary" conditions are satisfied. This is accomplished in Section 7.3.

DEFINITION 5.4.1. Suppose that $f : \mathbb{C} \to \mathbb{C}$ is a positively oriented map and $X \subset \mathbb{C}$ is a non-separating continuum. Suppose that there exist $n \geq 0$ disjoint non-separating continua Z_i such that the following properties hold:
 (1) $f(X) \setminus X \subset \cup_i Z_i$;
 (2) for all i, $Z_i \cap X = K_i$ is a non-separating continuum;
 (3) for all i, $f(K_i) \cap [Z_i \setminus K_i] = \emptyset$.
Then the map f is said to *scramble the boundary (of X)*. If instead of (3) we have
 (3a) for all i, either $f(K_i) \subset K_i$, or $f(K_i) \cap Z_i = \emptyset$
then we say that f *strongly scrambles the boundary (of X)*; clearly, if f strongly scrambles the boundary of X, then it scrambles the boundary of X. In either case, the continua K_i are called *exit continua (of X)*.

Observe that if in Definition 5.4.1 $n = 0$, then X must be invariant (i.e., $f(X) \subset X$).

REMARK 5.4.2. Since Z_i and $Z_i \cap X = K_i \neq \emptyset$ are non-separating continua and sets Z_i are pairwise disjoint, then $X \cup (\bigcup Z_i)$ is a non-separating continuum. Loosely, scrambling the boundary means that $f(X)$ can only "grow" off X *within* the sets Z_i and *through* the sets $K_i \subset X$ while any set K_i itself cannot be mapped outside X within Z_i, with more specific restrictions upon the dynamics of K_i's in the case of strong scrambling.

The following theorem extends Theorem 7.1.3 onto some non-invariant continua.

THEOREM 7.3.3. *In the situation of 5.4.1, if f is a positively oriented map which strongly scrambles the boundary of X, then f has a fixed point in X.*

We specify the above theorem for positively oriented maps with isolated fixed points as follows. Given a non-separating continuum $X \subset \mathbb{C}$, a positively oriented map f and a fixed point $p \in X$, we define what it means that f *repels outside X at p* (see Definition 7.4.5; basically, it means that there exists an invariant external ray to X which lands at p and along which the points are repelled away from p by f). We also need the next definition which is closely related to that of the index of the map on a simple closed curve.

DEFINITION 5.4.3. Suppose that $f : \mathbb{C} \to \mathbb{C}$ is a positively oriented map with isolated fixed points and x is a fixed point of f. Then the *local index of f at x*, denoted by $\mathrm{ind}(f, x)$, is defined as $\mathrm{ind}(f, S)$ where S is a small simple closed curve around x.

Then we prove the following theorem.

THEOREM 7.4.8. *Suppose that $f : \mathbb{C} \to \mathbb{C}$ is a positively oriented map with isolated fixed points, and $X \subset \mathbb{C}$ is a non-separating continuum or a point. Suppose that the conditions (1)-(3) in 5.4.1 are satisfied. Moreover, suppose that the following conditions hold.*

(1) *For each fixed point $p \in X$ we have that $\mathrm{ind}(f, p) = 1$ and f repels outside X at p.*
(2) *The map f scrambles the boundary of X. Moreover, for each i either $f(K_i) \cap Z_i = \emptyset$, or there exists a neighborhood U_i of K_i with $f(U_i \cap X) \subset X$.*

Then X is a point.

5.5. The polynomial case

Theorems 7.2.7 and 7.4.7 apply to polynomials acting on the complex plane. These theorems allow us to obtain corollaries dealing with the existence of periodic points in certain parts of the Julia set of a polynomial and with the degeneracy of certain continua (e.g., impressions). To discuss this we need the following standard notation.

Suppose that $P : \mathbb{C} \to \mathbb{C}$ is a complex polynomial of degree d. A P-periodic point a of period n is called *repelling* if $|(P^n)'(a)| > 1$, *parabolic* if $(P^n)'(a)$ is a root of unity (i.e., for an appropriate k we will have $[(P^n)'(a)]^k = 1$) and *irrational neutral* if $(p^n)'(a) = e^{2\pi\alpha i}$ with α irrational. The closure of the union of all repelling periodic points of P is called the *Julia set* of P and is denoted by J_P. Then the set $U^\infty(J_P) = U^\infty$ (i.e., the unbounded component of $\mathbb{C} \setminus J_P$) is called the *basin of attraction of infinity* and the set $K_P = \mathbb{C} \setminus U^\infty = T(J_P)$ is called the "filled-in" Julia set.

Components of $\mathbb{C} \setminus J_P$ are called *Fatou domains*. A Fatou domain is said to be *attracting (Siegel, respectively)* if it contains a periodic point which is attracting (irrational neutral, respectively); an irrational neutral periodic point like that is said to be a *Siegel (periodic) point*. A bounded periodic Fatou domain is said to be *parabolic* if it contains no periodic points (in this case all its points converge to the same parabolic periodic orbit which meets the boundary of the domain). Finally, an irrational neutral periodic point which belongs to J_P is said to be a *Cremer (periodic) point*.

The set U^∞ is foliated by so-called *(conformal) external rays* R_α of arguments $\alpha \in \mathbb{S}^1$. By [**DH85a**], if the degree of P is d and $\sigma_d : \mathbb{C} \to \mathbb{C}$ is defined by $\sigma_d(z) = z^d$, then $P(R_\alpha) = R_{\sigma_d(\alpha)}$. Denote by C_* the set of all preimages of critical points in $U^\infty(J_P)$ ($C_* = \emptyset$ if J_P is connected). If J_P is connected, each $\alpha \in \mathbb{S}^1$ corresponds to a unique external ray and all external rays are smooth and pairwise disjoint. In general R_α is smooth and unique if and only if $R_\alpha \cap C_* = \emptyset$. Other external rays are one-sided limits of smooth rays; it follows that they are non-smooth and there are at most countably many of them (in fact, for each $\alpha \in \mathbb{S}^1$ there exist at most two external rays $R_\alpha^\pm$ with argument α and each is a one sided limit of smooth external rays, see [**LP96**] for further details).

It is known that two distinct external rays are not homotopic in the complement of K_P (with the landing point fixed under the homotopies). Given an external ray R_α of K_P, we denote by $\Pi(R_\alpha) = \overline{R_\alpha} \setminus R_\alpha$ the *principal continuum of* R_α. Given a set $\mathfrak{R}$ of external rays, we extend the above notation by setting $\Pi(\mathfrak{R}) = \bigcup_{R \in \mathfrak{R}} \Pi(R)$. Now we are ready to give the following technical definition (see Figure 7.4 for an illustration).

DEFINITION 5.5.1 (General puzzle-piece). Let $P : \mathbb{C} \to \mathbb{C}$ be a polynomial. Let $X \subset K_P$ be a non-separating subcontinuum or a point such that the following holds.

(1) There exists $m \geq 0$ and m pairwise disjoint non-separating continua/points $E_1 \subset X, \ldots, E_m \subset X$.
(2) There exist m finite sets of external rays $A_1 = \{R_{a_1^1}, \ldots, R_{a_{i_1}^1}\}, \ldots, A_m = \{R_{a_1^m}, \ldots, R_{a_{i_m}^m}\}$ with $i_k \geq 2, 1 \leq k \leq m$.
(3) We have $\Pi(A_j) \subset E_j$ (so the set $E_j \cup (\bigcup_{k=1}^{i_j} R_{a_k^j}) = E_j'$ is closed and connected).
(4) X intersects a unique component C_X of $\mathbb{C} \setminus \cup E_j'$.
(5) For each Fatou domain U either $U \cap X = \emptyset$ or $U \subset X$.

We call such X a *general puzzle-piece* and call the continua E_i the *exit continua* of X. For each k, the set E_k' divides the plane into i_k open sets which we will call *wedges (at E_k)*; denote by W_k the wedge which contains $X \setminus E_k$ (it is well-defined by (4) above).

Note that if $m = 0$, $\bigcup E_j' = \emptyset$ and $C_X = \mathbb{C}$; so, any non-separating continuum in K_P with the empty set of exit continua satisfying (5) is a general puzzle-piece. Observe also, that there is a natural situation in which general puzzle-pieces can occur. Suppose that J_P is connected, conditions (1) - (3) are satisfied, and all continua E_j are contained in J_P while the continuum X is not yet defined. Suppose that there exists a component C of $\mathbb{C} \setminus \bigcup E_j'$ such that the boundary of C meets every $E_j, 1 \leq j \leq m$. Let $X = (C \cap K_P) \cup (\bigcup E_j)$. Then it is easy to see that X is a general puzzle-piece. However, our definition allows for a wider variety of general puzzle-pieces (like, e.g., non-separating invariant subcontinua of J_P).

For convenience call a fixed point x of a polynomial P *non-rotational* if there is a fixed external ray landing at x (it follows that each such point is either repelling or parabolic). We are ready to state the main result of Section 7.5.

THEOREM 7.5.2. Let P be a polynomial with filled-in Julia set K_P and let Y be a non-degenerate periodic component of K_P such that $P^p(Y) = Y$. Suppose that $X \subset Y$ is a non-degenerate general puzzle-piece with $m \geq 0$ exit continua $E_1, \ldots, E_m$ such that $P^p(X) \cap C_X \subset X$ and either $P^p(E_i) \subset W_i$, or E_i is a P^p-fixed point. Then at least one of the following claims holds:
  (1) X contains a P^p-invariant parabolic domain,
  (2) X contains a P^p-fixed point which is neither repelling nor parabolic, or
  (3) X has an external ray R landing at a repelling or parabolic P^p-fixed point such that $P^p(R) \cap R = \emptyset$ (i.e., P^p locally rotates at some parabolic or repelling P^p-fixed point).

Equivalently, suppose that Y is a non-degenerate periodic component of K_P such that $P^p(Y) = Y$, $X \subset Y$ is a general puzzle-piece with $m \geq 0$ exit continua $E_1, \ldots, E_m$ such that $P^p(X) \cap C_X \subset X$ and either $P^p(E_i) \subset W_i$, or E_i is a P^p-fixed point; if, moreover, X contains only non-rotational P^p-fixed points and does not contain P^p-invariant parabolic domains, then it is degenerate.

We also prove in Corollary 7.5.4 that an impression of an invariant external ray, to the filled in Julia set, which contains only repelling or parabolic periodic points is degenerate.

CHAPTER 6

Outchannels and their properties

6.1. Outchannels

In this section we will always let $f : \mathbb{C} \to \mathbb{C}$ be a continuous function. Suppose that X is a minimal continuum such that $f(X) \subset T(X)$ and f has no fixed point in $T(X)$. We show that X has at least one *negative outchannel*. We will always assume that (f, X, η) satisfies the standing hypothesis (see Definition 5.1.1 and the paragraph preceding 5.1.1) and see Section 4.4.1 for the notation $T(X)_\delta^\pm$. In particular, f is fixed point free on $T(X)_\eta$. Note that for each $\mathcal{KP}$-chord $\mathbf{g}$ in $T(X)_\eta$, $\mathrm{var}(f, \mathbf{g}, T(X)) = \mathrm{var}(f, \mathbf{g})$ is defined.

LEMMA 6.1.1. *Suppose that (f, X, η) satisfy the standing hypothesis and $\delta \leq \eta$. Let $Z \in \{T(X)_\delta^+, T(X)_\delta^-\}$. Fix a Riemann map $\varphi : \mathbb{D}^\infty \to \mathbb{C}^\infty \setminus Z$ such that $\varphi(\infty) = \infty$. Suppose R_t lands at $x \in \partial Z$. Then there is an open interval $M \subset \partial \mathbb{D}^\infty$ containing t such that φ can be extended continuously over M.*

PROOF. Suppose that $Z = T(X)_\delta^-$ and R_t lands on $x \in \partial Z$. By proposition 4.4.11 we may assume that $x \in X$. Note first that the family of chords in $\mathcal{KP}_\delta^-$ form a closed subset of the hyperspace of $\mathbb{C} \setminus X$, by Proposition 4.4.1. By symmetry, it suffices to show that we can extend ψ over an interval $[t', t] \subset \mathbb{S}^1$ for $t' < t$.

Let $\phi : \mathbb{D}^\infty \to \mathbb{C} \setminus T(X)$ be the Riemann map for $T(X)$. Then there exists $s \in \mathbb{S}^1$ so that the external ray R_s of $\mathbb{C} \setminus T(X)$ lands at x. Suppose first that there exists a chord $\mathbf{g} \in \mathcal{KP}_\delta^-$ such that $G = \varphi^{-1}(\mathbf{g})$ has endpoints s' and s with $s' < s$. Since $\mathcal{KP}_\delta^-$ is closed, there exists a minimal $s'' \leq s' < s$ such that there exists a chord $\mathbf{h} \in \mathcal{KP}_\delta^-$ so that $H = \varphi^{-1}(\mathbf{h})$ has endpoints s'' and s. Then $\mathbf{h} \subset \partial Z$ and ϕ can be extended over an interval $[t', t]$ for some $t' < t$, by Proposition 4.4.11 (4).

Suppose next that no such chord $\mathbf{g}$ exists. Choose a junction J_x for $T(X)_\delta^-$ and a neighborhood W of x such that $f(W) \cap [W \cup J_x] = \emptyset$. We will first show that there exists $\nu \leq \delta$ such that $x \in \partial T(X)_\nu$. For suppose that this is not the case. Then there exists a sequence $\mathbf{g}_i \in \mathcal{KP}$ of chords such that $x \in \mathrm{Sh}(\mathbf{g}_{i+1}) \subset \mathrm{Sh}(\mathbf{g}_i)$, $\lim \mathbf{g}_i = x$ and $\mathrm{var}(f, \mathbf{g}_i) > 0$ for all i. This contradicts Proposition 4.4.6. Hence $x \in \partial T(X)_\nu$ for some $\nu > 0$. We may assume that ν is so small that any chord of $\mathcal{KP}_\nu$ with endpoint x is contained in W.

By Proposition 4.4.12, the boundary of $T(X)_\nu$ is a simple closed curve S which must contain x. If there exists a chord $\mathbf{h} \in \mathcal{KP}_\nu$ with endpoint x such that H has endpoints s' and s with $s' < s$ then, since $\mathbf{h} \subset W$, $f(\mathbf{h}) \cap J_x = \emptyset$, $\mathrm{var}(f, \mathbf{h}) = 0$ and $\mathbf{h} \in \mathcal{KP}_\delta^-$, a contradiction. Similarly, all chords $\mathbf{h}$ close to x in S so that H has endpoints less than s and which are contained in W have $\mathrm{var}(f, \mathbf{h}) = 0$ by Proposition 4.4.6. Hence a small interval $[x', x] \subset S$, in the counterclockwise order on S is contained in $T(X)_\nu^-$. It now follows easily that a similar arc exists in the boundary of $T(X)_\delta^-$ and the desired result follows. □

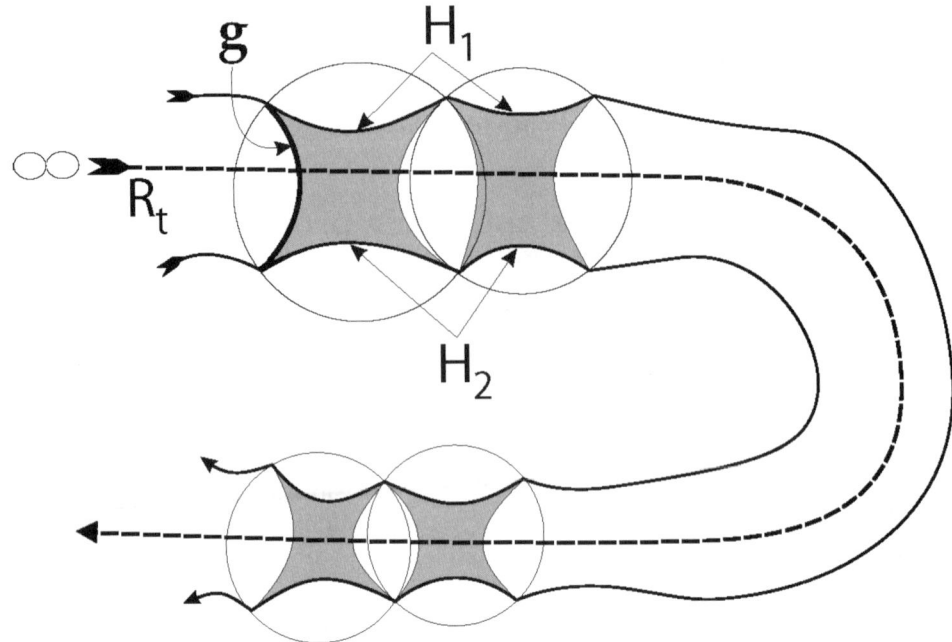

FIGURE 6.1. The strip $\mathfrak{S}$ from Lemma 6.1.2

By a *narrow strip* we mean the image of an embedding $h : \{(x,y) \in \mathbb{C} \mid x \geq 0 \text{ and } -1 < y < 1\} \to \mathbb{C}$ such that h has a continuous extension over the closure of its domain and $\lim_{x \to \infty} \operatorname{diam}(h(\{x\} \times [-1,1])) = 0$.

LEMMA 6.1.2. *Suppose that (f, X, η) satisfy the standing hypothesis. If there is a chord $\mathbf{g} \subset T(X)_\eta$ of $T(X)$ of negative (respectively, positive) variation, such that there is no fixed point in $T(T(X) \cup \mathbf{g})$, then there is a negative (respectively, positive) geometric outchannel $\mathcal{E}_t$ of $T(X)$ starting at $\mathbf{g}$.*

Moreover, if $\mathcal{E}_t$ is a positive (negative) geometric outchannel starting at the $\mathcal{KP}$-chord $\mathbf{g}$, and $\mathfrak{S} = \bigcup \{ \operatorname{conv}_\mathcal{H}(B \cap T(X)) \mid \operatorname{conv}_\mathcal{H}(B \cap T(X)) \subset T(X)_\eta \cap \overline{\operatorname{Sh}(\mathbf{g})} \text{ and a chord in } \operatorname{conv}_\mathcal{H}(B \cap T(X)) \text{ crosses } R_t \text{ essentially}\}$. Then $\mathfrak{S}$ is an infinite narrow strip in the plane whose remainder is contained in $T(X)$ and which is bordered by a $\mathcal{KP}$-chord and two halflines H_1 and H_2 (see figure 6.1).

PROOF. Without loss of generality, assume $\operatorname{var}(f, \mathbf{g}, T(X)) = \operatorname{var}(f, \mathbf{g}) < 0$. If $\mathbf{g}$ is such that for any chord $\mathbf{h} \subset T(X \cup \mathbf{g})$, $\mathbf{h} \subset T(X)_\eta$, put $\mathbf{g}' = \mathbf{g}$. Otherwise consider the boundary of $T(X)_\delta$ ($\delta < \eta$) which is locally connected by Proposition 4.4.12 and, hence, a Carathéodory loop. Then a continuous extension $g : \mathbb{S}^1 \to \partial T(X)_\delta$ of the Riemann map $\phi : \mathbb{D}^\infty \to \mathbb{C}^\infty \setminus T(X)_\delta$ exists. Whence the boundary of $T(X)_\delta$ contains a sub-path $A = g([a,b])$, which is contained in $\overline{\operatorname{Sh}(\mathbf{g})}$, whose endpoints coincide with the endpoints of $\mathbf{g}$. Note that for each component C of $A \setminus X$, $\operatorname{var}(f, C)$ is defined. Then it follows from Proposition 3.4.4, applied to a Carathéodory path, that there exists a component $C = \mathbf{g}'$ such that $\operatorname{var}(f, \mathbf{g}') < 0$. Note that $\mathbf{g}'$ is a $\mathcal{KP}$-chord contained in the boundary of $T(X)_\delta$. By taking δ sufficiently small we can assume that for any chord $\mathbf{h} \subset \overline{\operatorname{Sh}(\mathbf{g}')}$, $\mathbf{h} \subset T(X)_\eta$.

To see that a geometric outchannel, starting with $\mathbf{g}'$ exists, note that for any chord $\mathbf{g}'' \subset \overline{\mathrm{Sh}(\mathbf{g}')}$ with $\mathrm{var}(f, \mathbf{g}'', X) < 0$, if $\mathbf{g}'' = \lim \mathbf{g}_i$, then there exists i such that for any chord $\mathbf{h}$ which separates $\mathbf{g}_i$ and $\mathbf{g}''$ in $U^\infty(X)$, $\mathrm{var}(f, \mathbf{h}, X) = \mathrm{var}(f, \mathbf{g}'', X) < 0$. This follows since $f(\mathbf{g})$ is close to $f(\mathbf{h})$ and, hence, crosses a junction J_v in the same way (we can slightly change the junction J_v with $v \in \mathbf{g}''$ to a junction with vertex in $\mathbf{h}$ without changing the crossings of the images of the crosscuts with the junction). If $\mathbf{g}''$ is isolated on the side closest to X, then $\mathbf{g}'' \subset \mathrm{conv}_\mathcal{H}(B \cap T(X))$, where $\mathrm{conv}_\mathcal{H}(B \cap T(X))$ is a gap, such that $\mathbf{g}''$ separates $\mathrm{conv}_\mathcal{H}(B \cap T(X)) \setminus \mathbf{g}''$ from infinity in $U^\infty(X)$. Again by Proposition 3.4.4, there exists $\mathbf{h} \neq \mathbf{g}''$ in $\mathrm{conv}_\mathcal{H}(B \cap T(X))$ such that $\mathbf{g}''$ separates $\mathbf{h}$ from infinity in $U^\infty(X)$ and $\mathrm{var}(f, \mathbf{h}, X) < 0$. It follows from these two facts that there exists a maximal family of $\mathcal{KP}$ crosscuts, all of which have negative variation and are such that that for any three members of the family, one separates the other two in $U^\infty(X)$. Hence this maximal family determines a geometric outchannel. Each chord $\mathbf{h}$ in this family corresponds to a unique maximal ball $B_\mathbf{h}$. It is now not difficult to see that the union of all the sets $\mathrm{conv}_\mathcal{H}(B_\mathbf{h} \cap T(X))$ is a narrow strip. $\square$

6.1.1. Invariant Channel in X. We are now in a position to prove Bell's principal result on any possible counter-example to the fixed point property, under our standing hypothesis.

LEMMA 6.1.3. *Suppose $\mathcal{E}_t$ is a geometric outchannel of $T(X)$ under f. Then the principal continuum $\mathrm{Pr}(\mathcal{E}_t)$ of $\mathcal{E}_t$ is invariant under f. So $\mathrm{Pr}(\mathcal{E}_t) = X$.*

PROOF. Let $x \in \mathrm{Pr}(\mathcal{E}_t)$. Then for some chain $\{\mathbf{g}_i\}_{i=1}^\infty$ of crosscuts defining $\mathcal{E}_t$ selected from $\mathcal{KP}_\delta$, we may suppose $\mathbf{g}_i \to x \in \partial T(X)$ (by Lemma 4.4.8) and $\mathrm{var}(f, \mathbf{g}_i, X) \neq 0$ for each i. The external ray R_t meets all $\mathbf{g}_i$ and there is, for each i, a junction from $\mathbf{g}_i$ which "parallels" R_t. Since $\mathrm{var}(f, \mathbf{g}_i, X) \neq 0$, each $f(\mathbf{g}_i)$ intersects R_t. Since $\mathrm{diam}(f(\mathbf{g}_i)) \to 0$, we have $f(\mathbf{g}_i) \to f(x)$ and $f(x) \in \mathrm{Pr}(\mathcal{E}_t)$. We conclude that $\mathrm{Pr}(\mathcal{E}_t)$ is invariant. $\square$

THEOREM 6.1.4 (Dense channel, Bell). *If (X, f, η) satisfy our standing hypothesis then $T(X)$ contains a negative geometric outchannel; hence, $\partial U^\infty = \partial T(X) = X = f(X)$ is an indecomposable continuum.*

PROOF. By Lemma 4.4.12 $\partial T(X)_\eta$ is a Carathéodory loop. Since f is fixed point free on $T(X)_\eta$, $\mathrm{ind}(f, \partial T(X)_\eta) = 0$. Consequently, by Theorem 3.2.2 for Carathéodory loops, $\mathrm{var}(f, \partial T(X)_\eta) = -1$. By the summability of variation on $\partial T(X)_\eta$, it follows that on some chord $\mathbf{g} \subset \partial T(X)_\eta$, $\mathrm{var}(f, \mathbf{g}, T(X)) < 0$. By Lemma 6.1.2, there is a negative geometric outchannel $\mathcal{E}_t$ starting at $\mathbf{g}$.

Since $\mathrm{Pr}(\mathcal{E}_t)$ is invariant under f by Lemma 6.1.3, it follows that $\mathrm{Pr}(\mathcal{E}_t)$ is an invariant subcontinuum of $\partial U^\infty \subset \partial T(X) \subset X$. So by the minimality condition in our Standing Hypothesis, $\mathrm{Pr}(\mathcal{E}_t)$ is dense in X. It then follows from a theorem of Rutt [**Rut35**] that X is an indecomposable continuum. $\square$

THEOREM 6.1.5. *Assume that (X, f, η) satisfy our standing hypothesis and $\delta \leq \eta$. Then the boundary of $T(X)_\delta$ is a simple closed curve. The set of accessible points in the boundary of each of $T(X)_\delta^+$ and $T(X)_\delta^-$ is an at most countable union of pairwise disjoint continuous one-to-one images of $\mathbb{R}$.*

PROOF. By Theorem 6.1.4, X is indecomposable, so it has no cut points. By Proposition 4.4.12, $\partial T(X)_\delta$ is a Carathéodory loop. Since X has no cut points, neither does $T(X)_\delta$. A Carathéodory loop without cut points is a simple closed curve.

Let $Z \in \{T(X)_\delta^+, T(X)_\delta^-\}$ with $\delta \leq \eta$. Fix a Riemann map $\phi : \mathbb{D}^\infty \to \mathbb{C}^\infty \setminus Z$ such that $\phi(\infty) = \infty$. Corresponding to the choice of Z, let $\mathcal{W} \in \{\mathcal{KP}_\delta^+, \mathcal{KP}_\delta^-\}$. Apply Lemma 6.1.1 and find the maximal collection $\mathcal{J}$ of disjoint open subarcs of $\partial \mathbb{D}^\infty$ over which ϕ can be extended continuously. The collection $\mathcal{J}$ is countable. Since X has no cutpoints the extension is one-to-one over $\cup \mathcal{J}$. Since angles that correspond to accessible points are dense in $\partial \mathbb{D}^\infty$, so is $\cup \mathcal{J}$. If $Z = T(X)_\delta^+$, then it is possible that $\cup \mathcal{J}$ is all of $\partial \mathbb{D}^\infty$ except one point, but it cannot be all of $\partial \mathbb{D}^\infty$ since there is at least one negative geometric outchannel by Theorem 6.1.4. $\square$

Theorem 6.1.5 still leaves open the possibility that $Z \in \{T(X)_\delta^+, T(X)_\delta^-\}$ has a very complicated boundary. The set $C = \partial \mathbb{D}^\infty \setminus \cup \mathcal{J}$ is compact and zero-dimensional. Note that ϕ is discontinuous at points in C. We may call C the set of outchannels of Z. In principle, there could be an uncountable set of outchannels, each dense in X. The one-to-one continuous images of half lines in $\mathbb{R}$ lying in ∂Z are the "sides" of the outchannels. If two elements J_1 and J_2 of the collection $\mathcal{J}$ happen to share a common endpoint t, then the prime end $\mathcal{E}_t$ is an outchannel in Z, dense in X, with images of half lines $\phi(J_1)$ and $\phi(J_2)$ as its sides. It seems possible that an endpoint t of $J \in \mathcal{J}$ might have a sequence of elements J_i from $\mathcal{J}$ converging to it. Then the outchannel $\mathcal{E}_t$ would have only one (continuous) "side." Such exotic possibilities are eliminated in the next section.

In the proposition below we summarize several of the results in this section and show that an arc component K of the set of accessible points of the boundary of $T(X)_\delta^-$ is efficient in connecting close points in K. Note that it will follow later from Theorem 6.2.1 that there are no chords of positive variation. Hence $T(X)_\delta^- = T(X)_\delta$ which is always a simple closed curve.

PROPOSITION 6.1.6. Suppose that (X, f, η) satisfy our standing hypothesis, that the boundary of $T(X)_\delta^-$ is not a simple closed curve, $\delta \leq \eta$ and that K is an arc component of the boundary of $T(X)_\delta^-$ so that K contains an accessible point. Let $\varphi : \mathbb{D}^\infty \to \mathbb{C}^\infty \setminus T(X)_\delta^-$ be a conformal map such that $\varphi(\infty) = \infty$. Then:

(1) φ extends continuously and injectively to a map $\tilde{\varphi} : \tilde{\mathbb{D}}^\infty \to \tilde{U}^\infty$, where $\tilde{\mathbb{D}}^\infty \setminus \mathbb{D}^\infty$ is a dense and open subset of $\mathbb{S}^1$ which contains K in its image. Let $\tilde{\varphi}^{-1}(K) = (t', t) \subset \mathbb{S}^1$ with $t' < t$ in the counterclockwise order on $\mathbb{S}^1$. Hence $\tilde{\varphi}$ induces an order $<$ on K. If $x < y \in K$, we denote by $\langle x, y \rangle$ the subarc of K from x to y and by $\langle x, \infty \rangle = \cup_{y > x} \langle x, y \rangle$.

(2) $\mathcal{E}_t$ and $\mathcal{E}_{t'}$ are positive geometric outchannels of $T(X)$.

(3) Let R_t be the external ray of $T(X)_\delta^-$ with argument t. There exists $s \in R_t$, $B \in \mathfrak{B}^\infty$ and $\mathbf{g} \in \mathcal{KP}$ such that $s \in \mathbf{g} \subset \text{conv}_\mathcal{H}(B \cap X)$ and s is the last point of R_t in $\text{conv}_\mathcal{H}(B \cap X)$ (from ∞), $\mathbf{g}$ crosses R_t essentially and for each $B' \in \mathfrak{B}^\infty$ with $\text{conv}_\mathcal{H}(B' \cap X) \setminus X \subset \text{Sh}(\mathbf{g})$, $\text{diam}(B') < \delta$.

(4) There exists $\hat{x} \in K$ such that if $B' \in \mathfrak{B}^\infty$ with $\text{int}(B') \subset \text{Sh}(\mathbf{g})$, then $\text{conv}_\mathcal{H}(B' \cap X) \cap \langle \hat{x}, \infty \rangle$ is a compact ordered subset of K so that if C is $\mathcal{KP}$-crosscut in the boundary of $\text{conv}_\mathcal{H}(B' \cap X)$ with both endpoint in K, then $C \subset K$.

(5) Let $\mathfrak{B}_t^\infty \subset \mathfrak{B}^\infty$ be the collection of all $B \in \mathfrak{B}^\infty$ such that R_t crosses a chord in the boundary of $\text{conv}_\mathcal{H}(B \cap X)$ essentially and $\text{int}(B) \subset \text{Sh}(\mathbf{g})$. Then $\mathfrak{S} = \bigcup_{B \in \mathfrak{B}_t^\infty} \text{conv}_\mathcal{H}(B \cap X)$ is a narrow strip in the plane, bordered

by two halflines H_1 and H_2, which compactifies on X and one of H_1 or H_2 contains the set $\langle \hat{x}', \infty \rangle$ for some $\hat{x}' \in K$.

In particular, if $\max(\hat{x}, \hat{x}') < p < q$ and $\operatorname{diam}(\langle p, q \rangle) > 2\delta$, then there exists a chord $\mathbf{g} \in \mathcal{KP}$ such that one endpoint of $\mathbf{g}$ is in $\langle p, q \rangle$ and $\mathbf{g}$ crosses R_t essentially.

An analogous conclusion holds for $T(X)_\delta^+$ since its boundary cannot be a simple closed curve (clearly $T(X)_\delta$ must contain a crosscut of negative variation).

PROOF. By Proposition 4.4.11 and Theorem 6.1.5, and its proof, φ extends continuously and injectively to a map $\tilde{\varphi} : \tilde{\mathbb{D}}^\infty \to \tilde{U}^\infty$ and (1) holds.

By Lemma 6.1.1, the external ray R_t does not land. Hence there exist a chain $\mathbf{g}_i$ of $\mathcal{KP}_\delta$ chords which define the prime end $\mathcal{E}_t$. If for any i $\operatorname{var}(f, \mathbf{g}_i) \leq 0$, then $\mathbf{g}_i \subset T(X)_\delta^-$ a contradiction with the definition of t. Hence $\operatorname{var}(f, \mathbf{g}_i) > 0$ for all i sufficiently small and $\mathcal{E}_t$ is a positive geometric outchannel by the proof of Lemma 6.1.2. Hence (2) holds.

The proof of (3) is straightforward and is left to the reader.

Suppose that the endpoints of $\mathbf{g}$ are e and f with $f \in K$. Choose $\hat{x} > f$ in K so that $\hat{x}$ is the endpoint of a $\mathcal{KP}$ crosscut which is contained in $\operatorname{conv}_\mathcal{H}(B \cap X)$ with $B \subset \operatorname{Sh}(\mathbf{g})$. Let $B' \in \mathfrak{B}^\infty$ with $\operatorname{int}(B') \subset \operatorname{Sh}(\mathbf{g})$, $\hat{x} \notin B'$ and $\langle \hat{x}, \infty \rangle \setminus \operatorname{conv}_\mathcal{H}(B' \cap X)$ not connected. Suppose $\langle a, b \rangle$ is a bounded component of $\langle \hat{x}, \infty \rangle \setminus \operatorname{conv}_\mathcal{H}(B' \cap X)$ with endpoints in B'. Note that there must exist a chord $\mathbf{h} \in \mathcal{KP}$ with endpoints a and b. If $\operatorname{var}(f, \mathbf{h}) \leq 0$ we are done. Hence $\operatorname{var}(f, \mathbf{h}) > 0$. By Lemma 6.1.2, there is a geometric outchannel $\mathcal{E}_{t''}$ starting at $\mathbf{h}$. This outchannel disconnects the arc $\langle a, b \rangle$ between a and b, a contradiction. Hence (4) holds.

Next choose $\hat{x}' \in K$ such that each point of $\langle \hat{x}', \infty \rangle$ is accessible from $\operatorname{Sh}(\mathbf{g})$. Then each subarc $\langle p, q \rangle$ of $\langle \hat{x}, \infty \rangle$ of diameter bigger than 2δ cannot be contained in a single element of the $\mathcal{KPP}$ partition. Hence there exists a $\mathcal{KP}$-chord $\mathbf{g}$ which crosses R_t essentially and has one endpoint in $\langle p, q \rangle$.

Note that for each chord $\mathbf{h} \subset \operatorname{Sh}(\mathbf{g})$ which crosses R_t essentially, $\operatorname{var}(f, \mathbf{h}) > 0$. By Lemma 6.1.2, $\bigcup_{B \in \mathfrak{B}_t^\infty} \operatorname{conv}_\mathcal{H}(B \cap X)$ is a strip in the plane, bordered by two halflines H_1, H_2, which compactify on X. These two halflines, consist of chords in $\mathcal{KP}_\delta$ and points in X, one of which, say H_1 meets $\langle \hat{x}', \infty \rangle$. If $\langle \hat{x}', \infty \rangle$ is not contained in H_1 then, as in the proof of (4), there exists a chord $\mathbf{h} \subset H_1$ with $\operatorname{var}(f, \mathbf{h}) > 0$ joining two points of $x, y \in \langle \hat{x}, \infty \rangle$. As above this leads to a contradiction and the proof is complete. $\square$

6.2. Uniqueness of the Outchannel

Theorem 6.1.4 asserts the existence of at least one negative geometric outchannel which is dense in X. We show below that there is exactly one geometric outchannel, and that its variation is -1. Of course, X could have other dense channels, but they are "neutral" as far as variation is concerned.

THEOREM 6.2.1 (Unique Outchannel). If (X, f, η) satisfy the standing hypothesis then there exists a unique geometric outchannel $\mathcal{E}_t$ for X, which is dense in $X = \partial T(X)$. Moreover, for any sufficiently small chord $\mathbf{g}$ in any chain defining $\mathcal{E}_t$, $\operatorname{var}(f, \mathbf{g}, X) = -1$, and for any sufficiently small chord $\mathbf{g}'$ not crossing R_t essentially, $\operatorname{var}(f, \mathbf{g}', X) = 0$.

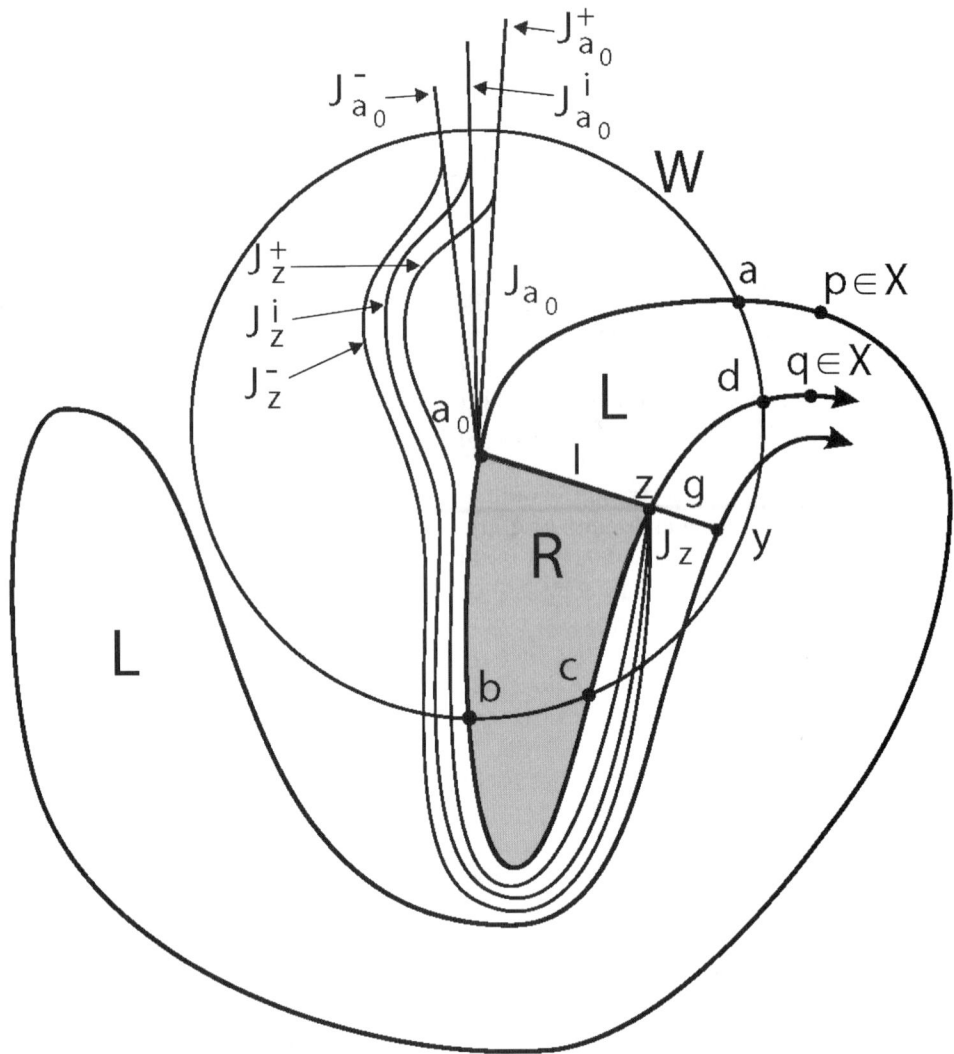

FIGURE 6.2. Uniqueness of the negative outchannel.

PROOF. Suppose by way of contradiction that X has a positive outchannel. Let $0 < \delta \leq \eta$ such that if $M \subset T(B(T(X), 2\delta))$ with $\operatorname{diam}(M) < 2\delta$, then $f(M) \cap M = \emptyset$. Since X has a positive outchannel, $\partial T(X)_\delta^-$ is not a simple closed curve. By Theorem 6.1.5 $\partial T(X)_\delta^-$ contains an arc component K which is the one-to-one continuous image of $\mathbb{R}$. Note that each point of K is accessible.

Let $\varphi : \mathbb{D}^\infty \to V^\infty = \mathbb{C} \setminus T(X)_\delta^-$ a conformal map. By Proposition 6.1.6, φ extends continuously and injectively to a map $\tilde{\varphi} : \tilde{\mathbb{D}}^\infty \to \tilde{V}^\infty$, where $\tilde{\mathbb{D}}^\infty \setminus \mathbb{D}^\infty$ is a dense and open subset of $\mathbb{S}^1$ which contains K in its image. Then $\tilde{\varphi}^{-1}(K) = (t', t) \subset \mathbb{S}^1$ is an open arc with $t' < t$ in the counterclockwise order on $\mathbb{S}^1$ (it could be that $(t', t) = \mathbb{S}^1 \setminus \{t\}$ and $t = t'$). By abuse of notation, let $<$ denote the order

6.2. UNIQUENESS OF THE OUTCHANNEL

in K induced by $\tilde{\varphi}$ and for $x < y$ in K, denote the arc in K with endpoints x and y by $\langle x, y \rangle$. For $x \in K$, let $\langle x, \infty \rangle = \cup_{y > x} \langle x, y \rangle$

Let $\mathcal{E}_t$ be the prime-end corresponding to t. By Proposition 6.1.6, $\Pr(\mathcal{E}_t)$ is a positive geometric outchannel and, hence, by Lemma 6.1.3, $\Pr(\mathcal{E}_t) = X$. Let $R_t = \varphi(re^{it})$, $r > 1$, be the external conformal ray corresponding to the prime-end $\mathcal{E}_t$ of $T(X)_\delta^-$. Since $\overline{R_t} \setminus R_t = X$ and the small chords $\mathbf{g}_x$ which define $\Pr(\mathcal{E}_t)$ have at least one endpoint in K, cross R_t essentially at x and have diameter going to zero if the endpoint in K moves in the positive direction along K, it follows that $X = \overline{R_t} \setminus R_t = \overline{\langle x, \infty \rangle} \setminus \langle x, \infty \rangle$.

By Proposition 6.1.6, there is $s \in R_t$ such that if $B \in \mathfrak{B}^\infty$ such that $\text{conv}_\mathcal{H}(B \cap X) \cap [(X, s)\text{-end of } R_t] \neq \emptyset$, then $\text{diam}(B) < \delta/2$. Let $\mathfrak{B}_t^\infty = \{B \in \mathfrak{B}^\infty \mid \text{conv}_\mathcal{H}(B \cap X) \text{ contains a chord } \mathbf{g} \text{ such that } \mathbf{g} \text{ crosses the } (X, s)\text{-end of } R_t \text{ essentially}\}$.

By Proposition 6.1.6 there exists $\hat{x} \in K \cap X$ such that for each arc $A \subset \langle \hat{x}, \infty \rangle$ with diameter $> 2\delta$, there is a $\mathcal{KP}$-chord $\mathbf{g}$ which contains a point of A as an endpoint and crosses R_t essentially.

Let $a_0 \in K \cap X$ so that $a_0 > \hat{x}$ and J_{a_0} is a junction of $T(X)_\delta^-$. Let W be an open disk, with simple closed curve boundary, about a_0 such that $\text{diam}(W) < \delta/4$ and $f(\overline{W}) \cap [\overline{W} \cup J_{a_0}] = \emptyset$. Let $a < a_0 < b$ in $K \cap \partial W$ such that $\langle a, b \rangle$ is the component of $K \cap \overline{W}$ which contains a_0. We may suppose that $\langle b, \infty \rangle \cap W$ is contained in one component of $W \setminus \langle a, b \rangle$ since one side of K is accessible from $\mathbb{C} \setminus T(X)_\delta^-$ and $a_0 \in X$. If $a \in X$, let $p = a$. If not, then there exists a $\mathcal{KP}$-chord $\mathbf{h} \subset K$ such that $a \in \mathbf{h}$. Let p be the endpoint of $\mathbf{h}$ such that $p < a$. See Figure 6.2.

Since $X \subset \overline{\langle x, \infty \rangle}$ there are components of $\langle b, \infty \rangle \cap W$ which are arbitrarily close to a_0. Choose $b < c < d$ in K so that the $\langle c, d \rangle$ is the closure of a component of $W \cap \langle b, \infty \rangle$ such that:

(1) a and d lie in the same component of $\partial W \setminus \{b, c\}$.
(2) There exists $z \in \langle c, d \rangle \cap X \cap W$ and an arc $I \subset \{a_0, z\} \cup [W \setminus \langle p, d \rangle]$ joining a_0 to z.
(3) There is a $\mathcal{KP}$-chord $\mathbf{g} \subset W$ with z and y as endpoints which crosses R_t essentially. Hence, $\text{var}(f, \mathbf{g}) > 0$.
(4) $\text{diam}(f(\mathbf{g})) < d(J_{a_0}^+ \setminus W, J_{a_0}^i \setminus W)$.

Conditions (1) and (2) follow because J_{a_0} is a connected and closed set from a_0 to ∞ in $\{a_0\} \cup [\mathbb{C} \setminus T(X)_\delta^-]$ and the ray $\langle b, \infty \rangle$ approaches both a_0 and p. Conditions (3) and (4) follow from Proposition 6.1.6. If $d \in X$, put $q = d$. Otherwise, let $q \in \langle d, \infty \rangle$, such that there is a $\mathcal{KP}$-chord $\mathbf{h} \subset K$ containing d with endpoint q.

By Corollary 3.4.2, there exists a bumping arc A' of $T(X)$ from p to q such that variation is defined on each component of $A' \setminus X$, $S' = A' \cup \langle p, q \rangle$ is a simple closed curve with $T(X) \subset T(S')$ and f is fixed point free on $T(S')$. Since $\overline{\mathbf{g}} \cap X = \{z, y\}$, we may assume that $A' \cap \overline{\mathbf{g}} = \{y\}$. Let C be the arc in ∂W from a to d disjoint from b. The arc A' may enter W and intersect I several times. However, in this case A' must enter W through C. Since we want to apply the Lollipop lemma, we will modify the arc A' to a new arc A which is disjoint from I.

Let A be the set of points in $A' \cup C$ accessible from ∞ in $\mathbb{C} \setminus [S' \cup C]$. Then A is a bumping arc from p to q, $A \cap I = \emptyset$, $\text{var}(f, A)$ is defined, $S = A \cup \langle p, q \rangle$ is a simple closed curve with $T(X) \subset T(S)$ and f is fixed point free on $T(S)$. Note that $y \in A$. Then the Lollipop lemma applies to S with $R = T(\langle a_0, z \rangle \cup I)$ and $L = T(I \cup \langle z, q \rangle \cup A \cup \langle p, a_0 \rangle)$.

Claim: $f(z) \in R$. Hence by Corollary 3.3.2, $\langle a_0, z \rangle$ contains a chord $\mathbf{g}_1$ with $\text{var}(f, \mathbf{g}_1) < 0$.

Proof of Claim 1. Note that the positive direction along $\mathbf{g}$ is from z to y. Since $z, y \in X$, $\{f(z), f(y)\} \subset X \subset T(S) = R \cup L$. Choose a junction J_z such that $J_{a_0} \setminus W \subset J_z$ and J_z runs close to $\langle a_0, z \rangle$ on its way to $\mathbf{g}$. In particular we may assume that $J_z \cap R = \{z\}$. Since $\mathbf{g}$ crosses R_t essentially, $\text{var}(f, \mathbf{g}) > 0$. For $* \in \{-, i, +\}$, let C_z^* be the union of components of $J_z^* \setminus W$ which are disjoint from $J_{a_0}^*$. Then C_z^i separates $R \cup C_z^+$ from $L \cup C_z^-$ in $\mathbb{C} \setminus W$ (see figure 6.2). Since $f(\mathbf{g}) \cap J_{a_0} = \emptyset$, if $f(z) \notin R$, $\text{var}(f, \mathbf{g}) \leq 0$, a contradiction. Hence $f(z) \in R$ (and, in fact, $f(y) \in L$) as desired.

Since $f(z) \in R$, $\langle a_0, z \rangle$ contains a chord $\mathbf{g}_1$ with $\text{var}(f, \mathbf{g}_1) < 0$. Repeating the same argument, replacing a_0 by z we obtain a second chord $\mathbf{g}_2$ contained in $\langle z, \infty \rangle$ such that $\text{var}(f, \mathbf{g}_2) < 0$.

We will now show that the existence of two distinct chords $\mathbf{g}_1$ and $\mathbf{g}_2$ in K with variation < 0 on each leads to a contradiction. Recall that $a_0 \in \overline{\langle b, \infty \rangle}$. Hence we can find $y' \in \langle b, \infty \rangle$ with $y' \in X$ such that $\mathbf{g}_1 \cup \mathbf{g}_2 \subset \langle a_0, y' \rangle$ and there exists a small arc $I' \subset W$ such that $I' \cap \langle a_0, y' \rangle = \{a_0, y'\}$. Since $f(I') \cap J_{a_0} = \emptyset$, $\text{var}(f, I') = 0$. We may also assume that f is fixed point free on $T(S'')$, where $S'' = I' \cup \langle a_0, y' \rangle$. Since $\langle a_0, y' \rangle$ contains both $\mathbf{g}_1$, $\mathbf{g}_2$ and no chords of positive variation, $\text{var}(f, \langle a_0, y' \rangle) \leq -2$ and $\text{var}(f, S'') \leq -2$. Then $\text{ind}(f, S'') = \text{var}(f, S'') + 1 \leq -1$ a contradiction with Theorem 3.1.4. Hence X has no positive geometric outchannel.

By Theorems 6.1.4 and 3.2.2, X has exactly one negative outchannel and its variation is -1. $\square$

Note that the following Theorem follows from Lemma 6.1.6 and Theorem 6.2.1.

THEOREM 6.2.2. *Suppose that X is a minimal counterexample to the Plane Fixed Point Problem. Then there exists $\delta > 0$ such that the continuum $Y = T(X)_\delta^+$ is a non-separating continuum, f is fixed point free on Y and all accessible points of Y are contained in one arc component K of the boundary of Y. In other words, Y is homeomorphic to a disk with exactly one channel removed which corresponds to the unique geometric outchannel of variation -1 of X. This channel compactifies on X. The sides of this channel are halflines consisting entirely of chords of zero variation and points in X. There exist arbitrarily small homeomorphisms of tails of these halflines to a tail of R_t which is the external ray corresponding to this channel.*

CHAPTER 7

Fixed points

In this chapter we study fixed points in invariant and non-invariant continua under positively oriented maps. We also obtain corollaries dealing with complex polynomials (the applications of these corollaries to complex dynamics are described in Chapter 1.)

7.1. Fixed points in invariant continua

In this section we will consider a positively oriented map of the plane. As we shall see below, a straight forward application of the tools developed above will give us the desired fixed point result. We will often assume, by way of contradiction, that $f : \mathbb{C} \to \mathbb{C}$ is a positively oriented map, X is a plane continuum such that $f(X) \subset T(X)$ and $T(X)$ contains no fixed points of f.

LEMMA 7.1.1. Let $f : \mathbb{C} \to \mathbb{C}$ be a map and $X \subset \mathbb{C}$ a continuum such that $f(X) \subset T(X)$. Suppose $C = (a, b)$ is a crosscut of the continuum $T(X)$. Let $v \in (a, b)$ be a point and J_v be a junction such that $J_v \cap (X \cup C) = \{v\}$. Then there exists an arc I such that $S = I \cup C$ is a simple closed curve, $T(X) \subset T(S)$ and $f(I) \cap J_v = \emptyset$.

PROOF. Since $f(X) \subset T(X)$ and $J_v \cap X = 0$, it is clear that there exists an arc I with endpoints a and b sufficiently close to $T(X)$ such that $I \cup C$ is a simple closed curve, $T(X) \subset T(I \cup C)$ and $f(I) \cap J_v = \emptyset$. This completes the proof. $\square$

COROLLARY 7.1.2. Suppose $X \subset \mathbb{C}$ is a continuum, $f : \mathbb{C} \to \mathbb{C}$ a positively oriented map such that $f(X) \subset T(X)$. Then for each crosscut C of $T(X)$ such that $f(\overline{C}) \cap \overline{C} = \emptyset$, $\text{var}(f, C) \geq 0$

PROOF. Suppose that $C = (a, b)$ is a crosscut of $T(X)$ such that $f(\overline{C}) \cap \overline{C} = \emptyset$ and $\text{var}(f, C) \neq 0$. Choose a junction J_v such that $J_v \cap (X \cup C) = \{v\}$ and $v \in C \setminus X$. By Lemma 7.1.1, there exists an arc I such that $S = I \cup C$ is a simple closed curve and $f(I) \cap J_v = \emptyset$. Moreover, by choosing I sufficiently close to X, we may assume that $v \in \mathbb{C} \setminus f(S)$. Hence $\text{var}(f, C) = \text{Win}(f, S, v) \geq 0$ by the remark following Definition 2.2.2. $\square$

THEOREM 7.1.3. Suppose $f : \mathbb{C} \to \mathbb{C}$ is a positively oriented map and X is a continuum such that $f(X) \subset T(X)$. Then there exists a point $x_0 \in T(X)$ such that $f(x_0) = x_0$.

PROOF. Suppose we are given a continuum X and $f : \mathbb{C} \to \mathbb{C}$ a positively oriented map such that $f(X) \subset T(X)$. Assume that $f|_{T(X)}$ is fixed point free. Choose a simple closed curve S such that $X \subset T(S)$ and points $a_0 < a_1 < \ldots < a_n$ in $S \cap X$ such that for each i $C_i = (a_i, a_{i+1})$ is a sufficiently small crosscut of X, $f(\overline{C_i}) \cap \overline{C_i} = \emptyset$ and $f|_{T(S)}$ is fixed point free. By Corollary 7.1.2, $\text{var}(f, C_i) \geq 0$

for each i. Hence by Theorem 3.2.2, $\mathrm{ind}(f, S) = \sum \mathrm{var}(f, C_i) + 1 \geq 1$. This contradiction with Theorem 3.1.4 completes the proof. $\square$

COROLLARY 7.1.4. *Suppose $f : \mathbb{C} \to \mathbb{C}$ is a perfect, oriented map and X is a continuum such that $f(X) \subset T(X)$. Then there exists a point $x_0 \in T(X)$ of period at most 2.*

PROOF. By Theorem 3.7.4, f is either positively or negatively oriented. In either case, the second iterate f^2 is positively oriented and must have a fixed point in $T(X)$ by Theorem 7.1.3. $\square$

7.2. Dendrites

Here we generalize Theorem 1.0.2 on the existence of fixed points in invariant dendrites to non-invariant dendrites. We also show that in certain cases the dendrite must contain infinitely many periodic cutpoints. Given two points a, b of a dendrite we denote by $[a, b], (a, b], [a, b), (a, b)$ the unique closed, semi-open and open arcs connecting a and b in the dendrite. Unless specified otherwise, the situation considered in this subsection is as follows: $D_1 \subset D_2$ are dendrites and $f : D_1 \to D_2$ is a continuous map. Set $E = \overline{D_2 \setminus D_1} \cap D_1$. In other words, E consists of points at which D_2 "grows" out of D_1. Observe that more than one component of $D_2 \setminus D_1$ may "grow" out of a point $e \in E$. We assume that D_1 is non-degenerate.

As an important tool we will need the following retraction closely related to the described above situation.

DEFINITION 7.2.1. For each $x \in D_2$ there exists a unique arc (possibly a point) $[x, d_x]$ such that $[x, d_x] \cap D_1 = \{d_x\}$. Hence there exists a natural monotone retraction $r : D_2 \to D_1$ defined by $r(x) = d_x$, and the map $g = g_f = r \circ f : D_1 \to D_1$ is a continuous map of D_1 into itself. We call the map r the *natural retraction (of D_2 onto D_1)* and the map g the *retracted (version of) f*.

The map g is designed to make D_1 invariant so that Theorem 1.0.2 applies to g and allows us to conclude that there are g-fixed points. Theorem 7.2.2 extends the result for $\mathbb{R}$ claiming that if there are points $a < b$ in $\mathbb{R}$ mapped by f in different directions, then there exists a fixed point $c \in (a, b)$ (see Introduction, Subsection 1.1). Let us recall the notion of the boundary scrambling property which is first introduced in Definition 5.3.1.

DEFINITION 5.3.1 (Boundary scrambling for dendrites). Suppose that f maps a dendrite D_1 to a dendrite $D_2 \supset D_1$. Put $E = \overline{D_2 \setminus D_1} \cap D_1$ (observe that E may be infinite). If for each *non-fixed* point $e \in E$, $f(e)$ is contained in a component of $D_2 \setminus \{e\}$ which intersects D_1, then we say that f has the *boundary scrambling property* or that it *scrambles the boundary*. Observe that if D_1 *is* invariant then f automatically scrambles the boundary.

We are ready to prove the following theorem.

THEOREM 7.2.2. *The following claims hold.*

(1) *If $a, b \in D_1$ are such that a separates $f(a)$ from b and b separates $f(b)$ from a, then there exists a fixed point $c \in (a, b)$. Thus, if $e_1 \neq e_2 \in E$ are such that $f(e_i)$ belongs to a component of $D_2 \setminus \{e_i\}$ disjoint from D_1 then there is a fixed point $c \in (e_1, e_2)$.*

(2) *If f scrambles the boundary, then f has a fixed point in D_1.*

Observe, that the fixed points found in (1) are cutpoints of D_1 (and hence of D_2).

PROOF. (1) Set $a_0 = a$. Then we find a sequence of points $a_{-1}, a_{-2}, \ldots$ in (a, b) such that $f(a_{-n-1}) = a_{-n}$ and a_{-n-1} separates a_{-n} from b. Clearly, $\lim_{n \to \infty} a_{-n} = c \in (a, b)$ is a fixed point as desired (by the assumptions c cannot be equal to b). If there are two points $e_1 \neq e_2 \in E$ such that $f(e_i)$ belongs to a component of $D_2 \setminus \{e_i\}$ disjoint from D_1 then the above applies to them.

(2) Assume that there are no f-fixed points $e \in E$. By Theorem 1.0.2 $g_f = g$ has a fixed point $p \in D_1$. It follows from the fact that f scrambles the boundary that points of E are not g-fixed. Hence $p \notin E$.

In general, a g-fixed point is not necessarily an f-fixed point. In fact, it follows from the construction that if $f(x) \neq g(x)$, then f maps x to a point belonging to a component of $D_2 \setminus D_1$ which "grows" out of D_1 at $r \circ f(x) = g(x) \in E$. Thus, since $g(p) = p$ but $p \notin E$, then $g(p) = f(p) = p$. □

REMARK 7.2.3. It follows from Theorem 7.2.2 that the only behavior of points in E which does not force the existence of a fixed point in D_1 is when one point $e \in E$ maps into a component of $D_2 \setminus \{e\}$ disjoint from D_1 whereas any other point $e' \in E$ maps into the component of $D_2 \setminus \{e'\}$ which is not disjoint from D_1.

Now we suggest conditions under which a map of a dendrite has infinitely many periodic cutpoints; the result will then apply in cases related to complex dynamics. Let us recall the notion of a weakly repelling periodic point which is first introduced in Definition 5.3.2.

DEFINITION 5.3.2 (Weakly repelling periodic points). In the situation of Definition 5.3.1, let $a \in D_1$ be a fixed point and suppose that there exists a component B of $D_1 \setminus \{a\}$ such that arbitrarily close to a in B there exist fixed cutpoints of D_1 or points x separating a from $f(x)$. Then we say that a is a *weakly repelling fixed point (of f in B)*. A periodic point $a \in D_1$ is said to be simply *weakly repelling* if there exists n and a component B of $D_1 \setminus \{a\}$ such that a is a weakly repelling fixed point of f^n in B.

Now we can prove Lemma 7.2.4.

LEMMA 7.2.4. Let a be a fixed point of f and B be a component of $D_1 \setminus \{a\}$. Then the following two claims are equivalent:
(1) a is a weakly repelling fixed point for f in B;
(2) either there exists a sequence of fixed cutpoints of $f|_B$, converging to a, or, otherwise, there exists a point $y \in B$ which separates a from $f(y)$ such that there are no fixed cutpoints in the component of $B \setminus \{y\}$ containing a in its closure (in the latter case for any $z \in (a, y]$ the point z separates $f(z)$ from a and each backward orbit of y in $(a, y]$ converges to a).

In particular, if a is a weakly repelling fixed point for f in B then a is a weakly repelling fixed point for f^n in B for any $n \geq 1$.

PROOF. Let us show that (2) implies (1). We may assume that there exists a point $y \in B$ which separates a from $f(y)$ such that there are no fixed cutpoints in the component W of $B \setminus \{y\}$ containing a in its closure. Choose a point $z \in (a, y)$. Since there are no fixed cutpoints of f in W, Theorem 7.2.2(1) implies that $f(z)$ cannot be separated from y by z. Hence $f(z)$ is separated from a by z, and a is weakly repelling for f in B. Moreover, we can take preimages of y in $(a, y]$, then take their preimages even closer to a, inductively. Any so constructed backward orbit of y in $(a, y]$ converges to a because it converges to a fixed point of f in $[a, y]$ and a is the only such fixed point.

Now, suppose that (1) holds. We may assume that there exists a neighborhood U of a in B such that there are no fixed cutpoints of f in U. If a is weakly repelling in B for f, we can choose a point $y \in U$ so that y separates a from $f(y)$ as desired.

It remains to prove the last claim of the lemma. Indeed, we may assume that there is no sequence of f^n-fixed cutpoints in B converging to a. Choose a neighborhood U of a which contains no f^n-fixed cutpoints in $U \cap B$. By (2) we can choose a point $y \in U \cap B$ such that y separates a from $f(y)$ so that there is a sequence of preimages of y under f which converges to a monotonically. Choosing the n-th preimage z we will see that z separates a from $f^n(z)$ with other parts of the second set of conditions of (2) also fulfilled. By the above a is weakly repelling for f^n in B as desired. □

Let B be a component of $D_1 \setminus \{a\}$ where a is fixed. Suppose that a is a weakly repelling fixed point for f in B which is not a limit of fixed cutpoints of f in B. Since the set of all vertices of D_2 together with their images under f and powers of f is countable (see Theorem 10.23 [**Nad92**]), we can choose y from Lemma 7.2.4 so that y and all cutpoints x in its backward orbit have $\mathrm{val}_{D_2}(x) = 2$. From now on to each fixed point a which is weakly repelling for f in a component B of $D_1 \setminus \{a\}$, but is not a limit point of fixed cutpoints in B, we associate a point $x_a \in B$ of valence 2 in D_2 separating a from $f(x_a)$ and such that all cutpoints in the backward orbit of x_a are of valence 2 in D_2. We also associate to a a semi-neighborhood U_a of a in $\overline{B}$ which is the component of $\overline{B} \setminus \{x_a\}$ containing a. We choose x_a so close to a that the diameter of U_a is less than one third of the diameter of B.

The next lemma shows that in some cases a fixed point p from Theorem 7.2.2(2) can be chosen to be a cutpoint of D_1. Recall that an endpoint of a continuum X is a point a such that the number $\mathrm{val}_X(a)$ of components of $X \setminus \{a\}$ equals 1.

LEMMA 7.2.5. *Suppose that f scrambles the boundary. Then either there is a fixed point of f which is a cutpoint of D_1, or, otherwise, there exists a fixed endpoint a of D_1 such that if C_a is the component of $D_2 \setminus \{a\}$ containing $D_1 \setminus \{a\}$, then a is not weakly repelling for f in C_a.*

PROOF. Suppose that f has no fixed cutpoints. By Theorem 7.2.2(2), the set of fixed points of f is not empty. Hence we may assume that *all* fixed points of f are endpoints of D_1 and, by way of contradiction, f is weakly repelling at any such fixed point a in the component of $D_2 \setminus \{a\}$ containing $D_1 \setminus \{a\}$. Suppose a and b are distinct fixed points of f. Let us show that either $U_a \subset U_b$, or $U_b \subset U_a$, or $U_a \cap U_b = \emptyset$. Set $\mathrm{diam}(D_1) = \varepsilon$.

Recall that x_a, x_b are cutpoints of D_2 of valence 2. Now, first we assume that $b \in U_a$. If $x_b \notin U_a$, then $U_a \subset U_b$ as desired. Suppose that $x_b \in U_a$. We will show that $x_b \in (b, x_a]$. Indeed, otherwise U_b would contain the component Q of $D_1 \setminus \{x_a\}$, not containing a. However, by the choice of the size of U_a we see that $\mathrm{diam}(Q) \geq 2\varepsilon/3$ and therefore $\mathrm{diam}(U_b) \geq 2\varepsilon/3$, a contradiction with the choice of the size of U_b. Hence $x_b \in (b, x_a]$ which implies that $U_b \subset U_a$. Now assume that $b \notin U_a$ and $a \notin U_b$. Then it follows that $x_a, x_b \in [a, b]$ and that the order of points in $[b, a]$ is b, x_b, x_a, a which implies that $U_b \cap U_a = \emptyset$.

Consider an open covering of the set of all fixed points $a \in D_1$ by their neighborhoods U_a and choose a finite subcover. By the above we may assume that it consists of pairwise disjoint sets $U_{a_1}, \ldots, U_{a_k}$. Consider the component Q of $D_1 \setminus \{x_{a_1}, \ldots, x_{a_k}\}$ whose endpoints are the points $x_{a_1}, \ldots, x_{a_k}$ and perhaps some endpoints of D_1. Then $f|_Q$ is fixed point free. On the other hand, $f|_{D_1}$ scrambles the boundary, and hence it is easy to see that $f|_Q$, with Q considered as a subdendrite of D_2, scrambles the boundary too, a contradiction with Theorem 7.2.2. □

Lemma 7.2.5 is helpful in the next theorem.

THEOREM 7.2.6. *Let D be a dendrite. Suppose that $f : D \to D$ is continuous and all its periodic points are weakly repelling. Then f has infinitely many periodic cutpoints.*

PROOF. By way of contradiction we assume that there are finitely many periodic cutpoints of f. Let us show that each endpoint b of D with $f(b) = b$ is a weakly repelling fixed point for f. Since the only component of $D \setminus \{b\}$ is $D \setminus \{b\}$, we will not be mentioning this component anymore. By the assumptions of the Theorem b is weakly repelling for some power f^m with $m \geq 1$. Then by Lemma 7.2.4 and by the assumption we can choose a point $x \neq b$ such that (1) x is not a vertex or endpoint of D, (2) for each point $z \in (b, x]$ we have that z separates b from $f^m(z)$, and (3) the component U of $D \setminus \{x\}$ containing b, contains no periodic cutpoints.

On the other hand, by way of contradiction we assume that b is not weakly repelling for f. Then, again by Lemma 7.2.4, no point $z \in (b, x]$ is such that z separates b from $f(z)$. The idea is to use this in order to find a point $y \in (b, x]$ which *does not* separate b from $f^m(y)$, a contradiction. To find y we apply the following construction. First, observe that there exists a point $d_1 \in (b, x)$ such that $f([b, x]) \supset [b, f(x)] \supset [b, d_1]$. Let $X_1 = \{z \in [b, x] \mid f(z) \in [b, x]\}$ be the set of points mapped into $[b, x]$ by f. Then $f(X_1) \supset [b, d_1]$ and all points of X_1 map towards b on $[b, x]$. We can apply the same observation to (b, d_1) instead of $(b, x]$. In this way we obtain a point $d_2 \in [b, d_1)$ and a set $X_2 = \{z \in [b, d_1] \mid f(z) \in X_1\}$ such that $[b, d_2] \subset f^2(X_2) \subset [b, d_1]$ and all points of $f(X_2)$ are mapped towards b by f. Repeating this argument, we will find points of $(b, x]$ mapped towards b and staying on $(b, x]$ for m steps in a row. This contradicts the previous paragraph and proves that if b is weakly repelling for f^m, then it is weakly repelling for f. Now by Lemma 7.2.4 b is weakly repelling for f^n for all $n \geq 1$.

Let f have finitely many periodic cutpoints $a^1, \ldots, a^k$ of f. For each a^i there exists N_i such that a^i is fixed for f^{N_i} and there exists a component B^i of $D \setminus \{a^i\}$ such that a^i is weakly repelling for f^{N_i} in B^i. Set $N = N_1 \cdots N_k$. Then it follows from Lemma 7.2.4 that each a^i is fixed for f^N and weakly repelling for f^N in B^i. Observe that, as we showed above, the endpoints of D which are fixed under f^N are in fact weakly repelling for f^N. Without loss of generality we may use f for f^N in the rest of the proof.

Let $A = \cup_{i=1}^k a^i$ and let B be a component of $D \setminus A$. Then $\overline{B}$ is a subdendrite of D to which the above tools apply: D plays the role of D_2, $\overline{B}$ plays the role of D_1, and E is exactly the boundary ∂B of B (by the construction $\partial B \subset A$). Suppose that each point $a \in \partial B$ is weakly repelling *in B*. Then all fixed points of f in B are endpoints of B, and all of them are weakly repelling for f. Thus, by Lemma 7.2.5 there exists a fixed cutpoint in B, a contradiction. Hence for some $a \in \partial B$ we have that a is *not* weakly repelling in $\overline{B} \setminus \{a\}$. By the assumption there exists a component, say, C', of $D \setminus \{a\}$ disjoint from B such that a *is* weakly repelling in C'. Let C be the component of $D \setminus A$ non-disjoint from C' with $a \in \partial C$.

We can now apply the same argument to C. If all boundary points of C are weakly repelling for f in C, then by Lemma 7.2.5 C will contain a fixed cutpoint, a contradiction. Hence there exists a point $d \in A$ such that d is *not* weakly repelling for f in C and a component F of $D \setminus A$ whose closure meets $\overline{C}$ at d, and d *is* weakly repelling in F. Note that $\overline{B} \cap \overline{F} = \emptyset$. Clearly, after finitely many steps this process will have to end (recall, that D is a dendrite), ultimately leading to a component Z of $D \setminus A$ such that all fixed points of f in $\overline{Z}$ are endpoints of $\overline{Z}$ at which f is weakly repelling. Again, Lemma 7.2.5 applies to $\overline{Z}$ and there exists a fixed cutpoint in $\overline{Z}$, a contradiction. □

An important application of Theorem 7.2.6 is to dendritic *topological Julia sets*. They can be defined as follows. Consider an equivalence relation $\sim$ on the unit circle $\mathbb{S}^1 \subset \mathbb{C}$. Equivalence classes of $\sim$ will be called *($\sim$-)classes* and will be denoted by boldface letters. A $\sim$-class consisting of two points is called a *leaf* ; a class consisting of at least three points is called a *gap* . Fix an integer $d > 1$ and define the map $\sigma_d : \mathbb{S}^1 \to \mathbb{S}^1$ by $\sigma_d(z) = z^d$, where z is a complex number with $|z| = 1$. Then the equivalence $\sim$ is said to be a *(d-)invariant lamination* (this is more restrictive than Thurston's definition in [**Thu09**]) if:

(E1) $\sim$ is *closed*: the graph of $\sim$ is a closed subset of $\mathbb{S}^1 \times \mathbb{S}^1$;

(E2) $\sim$ defines a *lamination*, i.e., it is *unlinked*: if $\mathbf{g}_1$ and $\mathbf{g}_2$ are distinct $\sim$-classes, then their convex hulls $\mathrm{Ch}(\mathbf{g}_1), \mathrm{Ch}(\mathbf{g}_2)$ in the unit disk $\mathbb{D}$ are disjoint,

(D1) $\sim$ is *forward invariant*: for a $\sim$-class $\mathbf{g}$, the set $\sigma_d(\mathbf{g})$ is a $\sim$-class too which implies that

(D2) $\sim$ is *backward invariant*: for a $\sim$-class $\mathbf{g}$, its preimage $\sigma_d^{-1}(\mathbf{g}) = \{x \in \mathbb{S}^1 : \sigma_d(x) \in \mathbf{g}\}$ is a union of $\sim$-classes;

(D3) for any gap $\mathbf{g}$, the map $\sigma_d|_{\mathbf{g}} : \mathbf{g} \to \sigma_d(\mathbf{g})$ is a *map with positive orientation*, i.e., for every connected component (s,t) of $\mathbb{S}^1 \setminus \mathbf{g}$ the arc $(\sigma_d(s), \sigma_d(t))$ is a connected component of $\mathbb{S}^1 \setminus \sigma_d(\mathbf{g})$.

The lamination in which all points of $\mathbb{S}^1$ are equivalent is said to be *degenerate*. It is easy to see that if a forward invariant lamination $\sim$ has a $\sim$-class with non-empty interior then $\sim$ is degenerate. Hence equivalence classes of any non-degenerate forward invariant lamination are totally disconnected.

Let $\sim$ define an invariant lamination. A $\sim$-class $\mathbf{g}$ is periodic if $\sigma^n(\mathbf{g}) = \mathbf{g}$ for some $n \geq 1$. Let $p : \mathbb{S}^1 \to J_\sim = \mathbb{S}^1/\sim$ be the quotient map of $\mathbb{S}^1$ onto its quotient space $J_\sim$. We can extend the equivalence relation $\sim$ to an equivalence relation $\approx$ of the entire plane by defining $x \approx y$ if either x and y are contained in the convex hull of one equivalence class of $\sim$, or $x = y$. Then the quotient map $m : \mathbb{C} \to \mathbb{C}/\approx$ is a monotone map whose point inverses are convex continua or points. Note that $p(\mathbb{S}^1) = \mathbb{S}^1/\sim = m(\overline{\mathbb{D}}) = \overline{\mathbb{D}}/\approx$. Let $f_\sim : J_\sim \to J_\sim$ be the map induced by σ_d. We call $J_\sim$ a *topological Julia set* and the induced map $f_\sim$ a *topological polynomial*. Recall that a *branched covering map* $f : X \to Y$ is a finite-to-one and open map for which there exists a finite set $F \subset Y$ such that $f|_{X \setminus f^{-1}(F)}$ is a covering map. Note that $f_\sim$ is a branched covering map, and in particular, $f_\sim$ has finitely many critical points (i.e., points where f is not locally one-to-one). It is easy to see that if $\mathbf{g}$ is a $\sim$-class then $\mathrm{val}_{J_\sim}(p(\mathbf{g})) = |\mathbf{g}|$ where by $|A|$ we denote the cardinality of a set A.

THEOREM 7.2.7. Suppose that the topological Julia set $J_\sim$ is a dendrite and $f_\sim : J_\sim \to J_\sim$ is a topological polynomial. Then all periodic points of $f_\sim$ are weakly repelling and $f_\sim$ has infinitely many periodic cutpoints.

PROOF. Suppose that x is an $f_\sim$-fixed point and set $\mathbf{g} = p^{-1}(x)$. Then $\sigma_d(\mathbf{g}) = \mathbf{g}$. Suppose first, that x is an endpoint of $J_\sim$. Then $\mathbf{g}$ is a singleton. Choose $y \neq x \in J_\sim$. Then the unique arc $[x, y] \subset J_\sim$ contains points $y_k \to x$ of valence 2 because there are no more than countably many vertices of $J_\sim$ (see Theorem 10.23 in [**Nad92**]). It follows that $\sim$-classes $p^{-1}(y_k)$ are leaves separating $\mathbf{g}$ from the rest of the circle and repelled from $\mathbf{g}$ under the action of σ_d which is expanding. Hence $f_\sim(y_i)$ is separated from x by y_i and so x is weakly repelling.

Suppose that x is not an endpoint. Choose a *very small* connected neighborhood U of x. It is easy to see that each component A of $U \setminus \{x\}$ corresponds to a single non-degenerate chord ℓ_A in the boundary of the Euclidean convex hull, $\mathrm{Ch}(\mathbf{g}) = G$, of $\mathbf{g}$. Recall that $\mathbb{S}^1 = \mathbb{R} \setminus \mathbb{Z}$ and that the endpoints a_A and b_A of ℓ_A are points in $\mathbb{S}^1$. Denote by $\sigma_d(\ell_A)$ the chord with endpoints $\sigma_d(a_A)$ and $\sigma_d(b_A)$ and by $|\ell_A| = \min\{|a_A - b_A|, 1 - |a_A - b_A|\}$, the *length of* ℓ_A. Since $f_\sim$ is a branched covering map, for each component A of $U \setminus \{x\}$ there exists a unique component $B = h(A)$ of $U \setminus \{x\}$ such that $f_\sim(A) \cap B \neq \emptyset$. This defines a map h from the set $\mathcal{A}$ of all components of $U \setminus \{x\}$ to itself. It follows that for each chord $\ell_A \subset \partial G$, $\sigma_d(\ell_A)$ is a non-degenerate chord in ∂G.

Suppose that there exist $\ell_A \subset \partial G$ and $n > 0$ such that $\sigma_d^n(\ell_A) = \ell_A$. Then it follows that the endpoints of ℓ_A are fixed under σ_d^{2n}. Connect x to a point $y \in A$ with the arc $[x, y]$, and choose, as in the first paragraph, a sequence of points $y_k \in [x, y], y_k \to x$ of valence 2. Then again by the expanding properties of σ_d^{2n} it follows that $f_\sim(y_i)$ is separated from x by y_i and so x is weakly repelling (for $f_\sim^{2n}$ in A).

It remains to show that there must exist a component A of $U \setminus \{x\}$ with $\sigma_d^n(\ell_A) = \ell_A$ for some $n > 0$. Clearly ∂G can contain at most finitely many chords ℓ_A such that its length $\mathrm{L}(\ell_A) \geq 1/(2(d+1))$. If $\mathrm{L}(\ell) < 1/(2(d+1))$, then $\mathrm{L}(\sigma_d(\ell)) = d \cdot \mathrm{L}(\ell)$ (i.e. σ_d expands the length of small leaves by the factor d).

Since the family of chords in the boundary of G is forward invariant and for each chord ℓ_A with $\mathrm{L}(\ell_A) < 1/(2(d+1))$, $\mathrm{L}(\sigma_d(\ell_A)) = d \cdot \mathrm{L}(\ell_A)$, such a periodic chord must exist (since if this is not the case there must exist an infinite number of distinct leaves in the boundary of G of length bigger than $1/(2(d+1))$, a contradiction.

Hence all periodic points of $f_\sim$ are weakly repelling and by Theorem 7.2.6 $f_\sim$ has infinitely many periodic cutpoints. □

7.3. Non-invariant continua and positively oriented maps of the plane

In this subsection we will extend Theorem 7.1.3 and obtain a general fixed point theorem which shows that if a non-separating plane continuum, not necessarily invariant, maps in an appropriate way, then it contains a fixed point. However we begin with Lemma 7.3.1 which gives a sufficient condition for the non-negativity of the variation of an arc.

LEMMA 7.3.1. *Let $f : \mathbb{C} \to \mathbb{C}$ be positively oriented, X a continuum and $C = [a, b]$ a bumping arc of X such that $f(a), f(b) \in X$ and $f(C) \cap C = \emptyset$. Let $v \in C \setminus X$, and let J_v be a junction at $v \in C$ defined as in Definition 2.2.2. If there exists a continuum K disjoint from J_v such that C is a bumping arc of K*

and $f(K) \cap J_v \subset \{v\}$ (e.g., K can be a subcontinuum of X with $a, b \in K$), then $\text{var}(f, C) \geq 0$.

PROOF. Consider two cases. First, let $f(K) \cap J_v = \emptyset$. Choose an arc I very close to K so that $S = I \cup C$ is a bumping simple closed curve around K and by continuity $f(I) \cap J_v = \emptyset$. Then $v \notin f(S)$. Moreover, since $f(I)$ is disjoint from J_v, it is easy to see that $\text{var}(f, C) = \text{win}(f, S, v) \geq 0$ (recall that f is positively oriented). Suppose now that $f(K) \cap J_v = \{v\}$. Then we can perturb the junction J_v slightly in a small neighborhood of v, obtaining a new junction J_d such that intersections of $f(C)$ with J_v and J_d are the same (and, hence, yield the same variation) and $f(K) \cap J_d = \emptyset$. Now proceed as in the first case. $\square$

Let us recall the notion of (strongly) scrambling the boundary of a planar continuum introduced in Definition 5.4.1 (it extends the notion of scrambling the boundary from maps of dendrites to maps of the plane).

DEFINITION 5.4.1. Suppose that $f : \mathbb{C} \to \mathbb{C}$ is a positively oriented map and $X \subset \mathbb{C}$ is a non-separating continuum. Suppose that there exist $n \geq 0$ disjoint non-separating continua Z_i such that the following properties hold:
 (1) $f(X) \setminus X \subset \cup_i Z_i$;
 (2) for all i, $Z_i \cap X = K_i$ is a non-separating continuum;
 (3) for all i, $f(K_i) \cap [Z_i \setminus K_i] = \emptyset$.
Then the map f is said to *scramble the boundary (of X)*. If instead of (3) we have
 (3a) for all i, either $f(K_i) \subset K_i$, or $f(K_i) \cap Z_i = \emptyset$
then we say that f *strongly scrambles the boundary (of X)*; clearly, if f strongly scrambles the boundary of X, then it scrambles the boundary of X. In either case, the continua K_i are called *exit continua (of X)*.

We will always use the same notation (for X, Z_i and K_i) introduced in Definition 5.4.1 unless explicitly stated otherwise. Let us make a few remarks. First, even though we use the notion only for positively oriented maps f, the definitions can be given for all continuous functions. Also, observe, that in the situation of Definition 5.4.1 if X is invariant then f automatically strongly scrambles the boundary because the set of exit continua can be taken to be empty. We will also agree that the choice of the sets Z_i is optimal in the sense that if $(f(X) \setminus X) \cap Z_i = \emptyset$ for some i, then the set Z_i will be removed from the list. In particular, all continua Z_i contain points from $f(X) \setminus X$ and hence all continua K_i have points from $\overline{f(X) \setminus X} \cap X$.

By Remark 5.4.2 $X \cup (\bigcup Z_i)$ is a non-separating continuum. Suppose that we are in the situation of the previous section, $D_1 \subset D_2$ are dendrites, $E = \overline{D_2 \setminus D_1} \cap D_1 = \{z_1, \ldots, z_l\}$ is finite, and $f : D_1 \to D_2$ scrambles the boundary in the sense of the previous section. For each z_i consider the union of all components of $D_2 \setminus D_1$ whose closures contain z_i, and denote by Z_i the closure of their union. In other words, Z_i is the closed connected piece of D_2 which "grows" out of D_1 at z_i. Then $Z_i \cap D_1 = \{z_i\}$, each Z_i is a dendrite itself, and f strongly scrambles the boundary in the sense of the new definition too. This explains why we use similar terminology in both cases.

From now on we fix a positively oriented map f. Even though some of the main applications of the results are to polynomial maps, this generality is well justified because in some arguments (e.g., when dealing with parabolic points) we have to locally perturb our map to make sure that the *local index* at a parabolic fixed point equals 1, and this leads to the loss of analytic properties of the map (see Lemma 7.5.1). Let us now prove the following technical lemma.

LEMMA 7.3.2. *Suppose that f is positively oriented, scrambles the boundary of X, Q is a bumping arc of X such that its endpoints map back into X and $f(Q) \cap Q = \emptyset$. Then $\mathrm{var}(f, Q) \geq 0$.*

PROOF. Suppose first that $Q \setminus \bigcup Z_i \neq \emptyset$ and choose $v \in Q \setminus \bigcup Z_i$. Since $v \in Q \setminus \bigcup Z_i$ and $X \cup (\bigcup Z_i)$ is non-separating, there exists a junction J_v, with $v \in Q$, such that $J_v \cap [X \cup Q \cup (\bigcup Z_i)] = \{v\}$ and, hence, $J_v \cap f(X) \subset \{v\}$. Now the desired result follows from Lemma 7.3.1.

Suppose now that $Q \setminus \bigcup Z_i = \emptyset$. Then $Q \subset Z_i$ for some i and so $Q \cap X \subset K_i$. In particular, both endpoints of Q belong to K_i. Choose a point $v \in Q$. Then again there is a junction connecting v and infinity outside X (except possibly for v). Since all sets $Z_j, j \neq i$ are positively distant from v and $X \cup (\bigcup_{i \neq j} Z_i)$ is non-separating, the junction J_v can be chosen to avoid all sets $Z_j, j \neq i$. Now, by part (3) of Definition 5.4.1, $f(K_i) \cap J_v \subset \{v\}$, hence by Lemma 7.3.1 $\mathrm{var}(f, Q) \geq 0$. □

Lemma 7.3.2 is applied in Theorem 7.3.3 in which we show that a map which strongly scrambles the boundary has fixed points. In fact, Lemma 7.3.2 is a major technical tool in our other results too. Indeed, suppose that a positively oriented map f scrambles the boundary of X. If we can construct a bumping simple closed curve S around X which has a partition into bumping arcs (*links of S*; see Definition 3.4.1) whose endpoints map into X (or at least into $T(S)$) and whose images are disjoint from themselves, then Lemma 7.3.2 would imply that the variation of S is non-negative. By Theorem 3.2.2 this would imply that the index of S is positive. Hence by Theorem 3.1.4 there are fixed points in $T(S)$. Choosing S to be sufficiently tight around X we see that there are fixed points in X. Thus, the construction of a tight bumping simple closed curve S with a partition satisfying the above listed properties becomes a major task.

For the sake of convenience we now sketch the proof of Theorem 7.3.3 which allows us to emphasize the main ideas rather than details. The main steps in constructing S are as follows. First we assume by way of contradiction that f has no fixed points in X. By Theorem 7.1.3 then $f(X) \not\subset X$ and $f(K_i) \not\subset K_i$ for any i. By the definition of strong scrambling then $f(K_i)$ is "far away" from Z_i for any i. Choose a tight bumping simple closed curve S around X with very small links. We need to construct a partition of S into bumping arcs whose endpoints map into X (or at least into $T(S)$) and whose images are disjoint from themselves. Since there are no fixed points in X, we may assume that all links of S move off themselves. However some of them may have endpoint(s) mapping outside X which prevents the corresponding partition from being the one we are looking for. So, we enlarge these links by consecutive concatenating them to each other until the images of the endpoints of these concatenations are inside X *and these concatenations still map off themselves* (the latter needs to be proven which is a big part of the proof of Theorem 7.3.3).

The bumping simple closed curve S then remains as before, but the partition changes because we enlarge some links. Still, the construction shows that the new partition is satisfactory, and since S can be chosen arbitrarily tight, this implies the existence of a fixed point in X as explained before. Thus, a new development is that we are able to construct a partition of S which has all the above listed necessary properties having possibly *very long* links.

To achieve the goal of replacing some links in S by their concatenations we consider the links with at least one endpoint mapped outside X in detail (indeed, Lemma 7.3.2 already applies to all other links) and use the fact that f strongly scrambles the boundary. The idea is to consider consecutive links of S with endpoints mapped into $Z_i \setminus X$. Their concatenation is a connected piece of S with endpoints (and a lot of other points) belonging to X and mapping into one Z_i. If we begin the concatenation right before the images of links enter $Z_i \setminus X$ and stop it right after the images of the links exit $Z_i \setminus X$ we will have one condition of Lemma 7.3.2 satisfied because the endpoints of the thus constructed new "big" concatenation link T of S map into X.

We need to verify that T moves off itself under f. This is easy to see for the end-links of T: each end-link has the image "crossing" into X from $Z_i \setminus X$, hence the images of end-links are close to K_i. However the set K_i is mapped "far away" from Z_i by the definition of strong scrambling and because none of the K_j's is invariant by the assumption. This implies that the end-links themselves must be far away from K_i (and hence from Z_i). If now we move from link to link inside T we see that those links cannot approach Z_i too closely because if they do, they will have to "be close to K_i", and their images will have to be close to the image of K_i which is far away from Z_i, a contradiction with the fact that all links in T have endpoints which map into $Z_i \setminus X$. In other words, the dynamics of K_i prevents the new bigger links from getting even close to Z_i under f which shows that T moves off itself as desired (after all, the images of new bigger links are close to the set $Z_i \setminus X$).

Given a compact set K denote by $B(K, \varepsilon)$ the open set of all points whose distance to K is less than ε. By $d(\cdot, \cdot)$ we denote the distance between two points or sets.

THEOREM 7.3.3. *Suppose $f : \mathbb{C} \to \mathbb{C}$ is positively oriented, X is a non-separating continuum and f strongly scrambles the boundary of X. Then f has a fixed point in X.*

PROOF. If $f(X) \subset X$ then the result follows from Theorem 7.1.3. Similarly, if there exists i such that $f(K_i) \subset K_i$, then f has a fixed point in $K_i \subset X$ and we are also done. Hence we may assume that $f(X) \setminus X \neq \emptyset$, there are $m > 0$ sets $Z_i, i = 1, \ldots, m$, $(f(X) \setminus X) \cap Z_i \neq \emptyset$ for any i, and $f(K_i) \cap Z_i = \emptyset$ for all i (making these claims we rely upon the fact that f strongly scrambles the boundary). Suppose that $f|_X$ is fixed point free. Then there exists $\varepsilon > 0$ such that for all $x \in X$, $d(x, f(x)) > \varepsilon$. We may assume that $2\varepsilon < \min\{d(Z_i, Z_j) \mid i \neq j\}$. We now choose constants η', η, δ and a bumping simple closed curve S (whose initial links are crosscuts) of X so that the following holds.

(1) $0 < \eta' < \eta < \delta < \varepsilon/3$.
(2) For each $x \in X \cap B(K_i, 3\delta)$ we have $d(f(x), Z_i) > 3\delta$.
(3) For each $x \in X \setminus B(K_i, 3\delta)$ we have $d(x, Z_i) > 3\eta$.
(4) For each i there is a point $x_i \in X$ with $f(x_i) = z_i \in Z_i$ and $d(z_i, X) > 3\eta$. Since by Theorem 3.7.4 $\partial f(X) \subset f(\partial X)$ and X is non-separating, we may assume that $x_i \in \partial X$.
(5) $X \subset T(S)$ and $A = X \cap S = \{a_0 < \cdots < a_n < a_{n+1} = a_0\}$ with points of A numbered in the positive circular order around S.

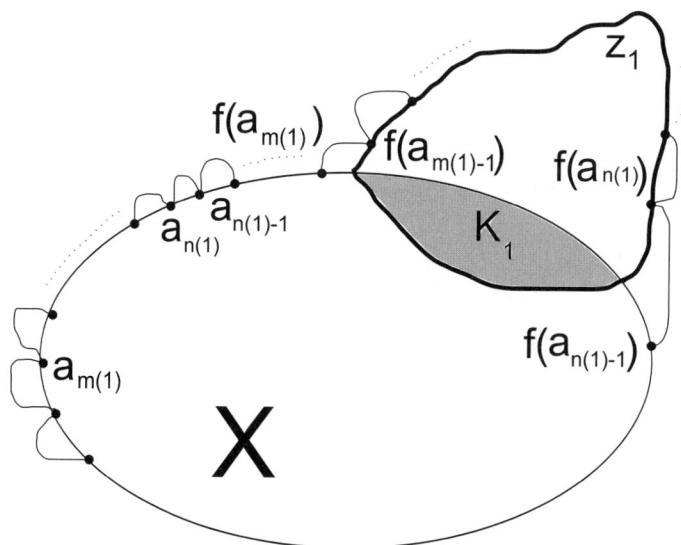

FIGURE 7.1. Replacing the links $[a_{n(1)-1}, a_{n(1)}]$, ..., $[a_{m(1)-1}, a_{m(1)}]$ by a single link $[a_{n(1)-1}, a_{m(1)}]$.

(6) $f|_{T(S)}$ is fixed point free.
(7) For any $Q_i = (a_i, a_{i+1}) \subset S$, $\mathrm{diam}(\overline{Q_i}) + \mathrm{diam}(f(\overline{Q_i})) < \eta$.
(8) For any $x, y \in X$ with $d(x,y) < \eta'$ we have $d(f(x), f(y)) < \eta$.
(9) A is an η'-net in ∂X (i.e., the Hausdorff distance between A and ∂X is less than η').

Observe that $\overline{Q_i}$ is contained in the closed ball centered at a_i of radius $\mathrm{diam}(\overline{Q_i})$ and $f(\overline{Q_i})$ is contained in the closed ball centered at $f(a_i)$ of radius $\mathrm{diam}(f(\overline{Q_i}))$; hence by (7) and since $d(x, f(x)) > \varepsilon$ for all $x \in X$ we see that $\overline{Q_i} \cap f(\overline{Q_i}) = \emptyset$ for every i (we rely on the triangle inequality here too).

Claim 1. *There exists a point $a_j \in A$ such that $f(a_j) \in X \setminus \bigcup_i \overline{B(Z_i, \eta)}$.*

Proof of Claim 1. Set $\overline{B(Z_i, 3\eta)} = T_i$. We will show that there exists a point $x \in \partial X$ with $f(x) \in X \setminus \bigcup T_i$. Indeed, suppose first that $m = 1$. Then by (2) and the assumption that $f(K_i) \cap Z_i = \emptyset$ for each i we have $f(K_1) \subset X \setminus T_1$, and we can choose any point of $K_1 \cap \partial X$ as x. Now, suppose that $m \geq 2$. Observe that by the choice of ε and by (1) the compacta T_i are pairwise disjoint. By (4) for each i there are points x_i in ∂X such that $f(x_i) \in Z_i \subset T_i$. Since the sets $f^{-1}(T_i) \cap X$ are pairwise disjoint non-empty compacta we see that the set $V = \partial X \setminus \bigcup f^{-1}(T_i)$ is non-empty (because ∂X is a continuum). Now we can choose any point of V as x.

Notice now that by the choice of A (see (9)) we can find a point a_j such that $d(a_j, x) < \eta'$ which by (8) implies that $d(f(a_j), f(x)) < \eta$ and hence $f(a_j) \in X \setminus (\bigcup \overline{B(Z_i, \eta)})$ as desired. □

By Claim 1, we assume without loss of generality, that $f(a_0) \in X \setminus \bigcup \overline{B(Z_i, \eta)}$. Now, by (4) there exists a point x_1 such that $f(x_1) = z_1$ is more than 3η-distant from X. We can find $a_l \in A$ such that $d(a_l, x_1) < \eta'$ and hence by (8) and by the triangle inequality $f(a_l)$ is at least 2η-distant from X. On the other hand, $f(a_0) \in X$. By the choice of a_0, we can find minimal $n(1) < m(1)$ from $\{0, 1, \ldots, m+1\}$ such that the following claims hold. Without loss of generality, $n(1) = 1$.

(1) $f(a_{n(1)-1}) \in X$.
(2) $f(a_r) \in f(X) \setminus X$ for all r with $n(1) \leq r \leq m(1) - 1$ (and so, since $\operatorname{diam}(f(Q_u)) < \varepsilon/3$ for any u and $d(Z_s, Z_t) > 2\varepsilon$ for all $s \neq t$, there exists $i(1) \in \{1, \ldots, n\}$ with $f(a_r) \in Z_{i(1)}$ for all $n(1) \leq r \leq m(1) - 1$).
(3) $f(a_{m(1)}) \in X$.

Consider the arc $Q' = [a_{n(1)-1}, a_{m(1)}] \subset S$ and show that $f(Q') \cap Q' = \emptyset$. As we walk along Q' and mark the f-images of points $a_{n(1)-1}, a_{n(1)}, \ldots, a_{m(1)}$, we begin in X at $f(a_{n(1)-1})$, then enter $Z_{i(1)} \setminus X$ and walk inside it, and then exit $Z_{i(1)} \setminus X$ at $f(a_{m(1)}) \in X$. Since every step in this walk is rather short (by (7) $\operatorname{diam}(Q_i) + \operatorname{diam}(f(Q_i)) < \eta$), we see that $d(f(a_{n(1)-1}), Z_{i(1)} \setminus X) < \eta$ and $d(f(a_{m(1)}), Z_{i(1)} \setminus X) < \eta$. On the other hand for each $r, n(1) \leq r \leq m(1) - 1$, we have $f(a_r) \in Z_{i(1)} \setminus X$. Thus, $d(f(a_r), Z_{i(1)}) < \eta$ for each $n(1) - 1 \leq r \leq m(1)$. Since by (7) for each link Q of S we have $\operatorname{diam}(Q) + \operatorname{diam}(f(Q)) < \eta$, we now see by the triangle inequality that $d(f(Q'), Z_{i(1)}) < 2\eta$.

This implies that for $n(1) - 1 \leq r \leq m(1)$, $d(a_r, K_{i(1)}) > 3\delta$ (because otherwise by (2) $f(a_r)$ would be farther away from $Z_{i(1)}$ than $3\delta > \eta$, a contradiction) and so $d(a_r, Z_{i(1)}) > 3\eta$ (because $a_r \in X \setminus B(K_{i(1)}, 3\delta)$ and by (3)). Since by (7) for each link of S we have $\operatorname{diam}(Q) + \operatorname{diam}(f(Q)) < \eta$, then $d(Q', Z_{i(1)}) > 2\eta$.

Therefore $f(Q') \cap Q' = \emptyset$. This allows us to replace the original division of S into links $Q_0, \ldots, Q_{m(1)-1}$ by a new one in which Q' plays the role of a new link; in other words, we simply delete the points $\{a_{n(1)}, \ldots, a_{m(1)-1}\}$ from A. Thus, Q' is a bumping arc whose endpoints map back into the continuum X and such that $f(Q') \cap Q' = \emptyset$. Therefore Q' satisfies the conditions of Lemma 7.3.2, and so $\operatorname{var}(f, Q') \geq 0$. Observe also that for Q' the associated continuum $Z_{i(1)}$ is well-defined because the distance between distinct continua Z_i is greater than 2ε. Replace the string of links $\{Q_0, \ldots, Q_{m(1)-1}\}$ in S by the single link $Q' = Q'_0$ which has as endpoints $a_{n(1)-1}$ and $a_{m(1)}$. Continuing in the same manner and moving along S, in the end we obtain a finite set $A' = \{a_0 = a'_0 < a'_1 < \cdots < a'_k\} \subset A$ such that for each i we have $f(a'_i) \in X \subset T(S)$ and for each arc $Q'_i = [a'_i, a'_{i+1}]$ we have $f(Q'_i) \cap Q'_i = \emptyset$. In other words, we will construct a partition of S satisfying all the required properties: its links are bumping arcs whose endpoints map back into X and whose images are disjoint from themselves. As outlined after Lemma 7.3.2, this yields a contradiction. More precisely, by Theorem 3.2.2, $\operatorname{ind}(f, S) = \sum \operatorname{var}(f, Q'_i) + 1$, and since by Lemma 7.3.2, $\operatorname{var}(f, Q'_i) \geq 0$ for all i, $\operatorname{ind}(f, S) \geq 1$ contradicting the fact that f is fixed point free in $T(S)$ (see Theorem 3.1.4). $\square$

7.4. Maps with isolated fixed points

In this section we assume that all maps $f : \mathbb{C} \to \mathbb{C}$ are positively oriented maps with isolated fixed points.

DEFINITION 7.4.1. Given a map $f : X \to Y$ we say that $c \in X$ is a *critical point* of f if for each neighborhood U of c, there exist $x_1 \neq x_2 \in U$ such that $f(x_1) = f(x_2)$. Hence, if x is not a critical point of f, then f is locally one-to-one near x.

If a point x belongs to a non-degenerate continuum collapsed to a point under f then x is critical; also any point which is an accumulation point of collapsing continua is critical. However in these cases the map near x may be monotone. A more interesting case is when the map near x is not monotone; then x is a *branchpoint* of f and it is critical even if there are no collapsing continua close by. One can define the *local degree* $\deg_f(a)$ as the number of components of $f^{-1}(y)$ non-disjoint from a small neighborhood of a (y then should be chosen close to $f(a)$). It is well-known that for a positively oriented map f and a point a which is a component of $f^{-1}(f(a))$ the local degree $\deg_f(a)$ equals the winding number $\text{win}(f, S, f(a))$ for any small simple closed curve S around a. Then branchpoints are exactly the points at which the local degree is greater than 1. Notice that since we do not assume any smoothness, a *critical* point may well be both fixed (periodic) and topologically repelling in the sense that some small neighborhoods of $c = f(c)$ map over themselves by f.

Let us recall the notion of a local index of a map at a point which is first introduced in Definition 5.4.3.

DEFINITION 5.4.3. Suppose that $f : \mathbb{C} \to \mathbb{C}$ is a positively oriented map with isolated fixed points and x is a fixed point of f. Then the *local index of f at x*, denoted by $\text{ind}(f, x)$, is defined as $\text{ind}(f, S)$ where S is a small simple closed curve around x.

It is easy to see that, since f is positively oriented and has isolated fixed points, the local index is well-defined, i.e. does not depend on the choice of S. By modifying a translation map one can give an example of a homeomorphism of the plane which has exactly one fixed point x with local index 0. Still in some cases the local index at a fixed point must be positive.

DEFINITION 7.4.2. Let $f : \mathbb{C} \to \mathbb{C}$ be a map. A fixed point x is said to be *topologically repelling* if there exists a sequence of simple closed curves $S_j \to \{x\}$ such that $x \in \text{int}(T(S_j)) \subset T(S_j) \subset \text{int}(T(f(S_j)))$. A fixed point x is said to be *topologically attracting* if there exists a sequence of simple closed curves $S_j \to \{x\}$ not containing x and such that $x \in \text{int}(T(f(S_j))) \subset T(f(S_j)) \subset \text{int}(T(S_j))$.

LEMMA 7.4.3. Let $f : \mathbb{C} \to \mathbb{C}$ be a positively oriented map with isolated fixed points. If a is a topologically repelling fixed point then we have that $\text{ind}(f, a) = \deg_f(a) \geq 1$. If however a is a topologically attracting fixed point then $\text{ind}(f, a) = 1$.

PROOF. Consider the case of the repelling fixed point a. Then it follows that, as x runs along a small simple closed curve S with $a \in T(S)$, the vector from x to $f(x)$ produces the same winding number as the vector from a to $f(x)$. As we remarked before, it is well-known that this winding number equals $\deg_f(a)$; on the other hand, $\text{ind}(f, S) > 0$ since f is positively oriented and has isolated fixed points. The argument for an attracting fixed point is similar. $\square$

If however a fixed point x is neither topologically repelling nor topologically attracting, then $\text{ind}(f, x)$ could be greater than 1 even in the non-critical case. Indeed, by definition $\text{ind}(f, x)$ coincides with the winding number of $f(z) - z$ on a small simple closed curve S around x with respect to the origin. If, e.g., f is rational and $f'(x) \neq 1$ then this implies that $\text{ind}(f, x) = 1$. However if $f'(x) = 1$ then $\text{ind}(f, x)$ is the multiplicity at x (i.e., the local degree of the map $f(z) - z$

at x). This is related to the following useful theorem. It is a version of the more general topological *argument principle*.

THEOREM 7.4.4. *Suppose that f is positively oriented and has isolated fixed points. Then for any simple closed curve $S \subset \mathbb{C}$, which contains no fixed points of f, its index equals the sum of local indices taken over all fixed points in $T(S)$. In particular if for each fixed point $p \in T(S)$ we have that $\text{ind}(f,x) = 1$ then $\text{ind}(f,S)$ equals the number $n(f,S)$ of fixed points inside $T(S)$.*

Theorem 7.4.4 implies Theorem 3.1.4 for positively oriented maps with isolated fixed points (indeed, if $\text{ind}(f,S) \neq 0$ then by Theorem 7.4.4 there must exist fixed points in $T(S)$), and actually provides more information. By the above analysis, Lemma 7.4.3 and Theorem 7.4.4, $\text{ind}(f,S)$ equals the number $n(f,S)$ of fixed points inside $T(S)$ if all f-fixed points in $T(S)$ are either topologically attracting, or such that f has a complex derivative f' at x, and $f'(x) \neq 1$; if f-fixed points can also be topologically repelling, then $\text{ind}(f,S) \geq n(f,S)$

In the spirit of the previous parts of the paper, we are still concerned with finding f-fixed points inside non-invariant continua of which f (strongly) scrambles the boundary. However we now specify the types of fixed points we are looking for. Thus, the main result of this subsection proves the existence of specific fixed points in non-degenerate continua satisfying the appropriate boundary conditions and shows that in some cases such continua must be degenerate. It is in this form that we apply the result later in this subsection.

Recall that an essential crossing of an external ray R and a crosscut Q was defined in Definition 3.6.4; there an external ray R_t is said to *cross* a crosscut Q *essentially* if and only if there exists $x \in R_t$ such that the bounded component of $R_t \setminus \{x\}$ is contained in the bounded complementary domain of $T(X) \cup Q$. The fact that a crosscut crosses a ray essentially can be similarly restated in the language of the uniformization plane (i.e., if the ray and the crosscut are replaced by their counterparts in the uniformization plane while X is replaced by the unit disk in the uniformization plane).

For the next definition we need to make an observation. Suppose that $f : \mathbb{C} \to \mathbb{C}$ is a map and D is a closed Jordan disk with interior non-disjoint from a continuum X such that $f(D \cap X) \subset X$ and $f(\overline{\partial D \setminus X}) \cap D = \emptyset$. Suppose in addition that $|\partial D \cap X| \geq 2$. Then the closure of any component Q of $\partial D \setminus X$ is a bumping arc whose endpoints map back into X and such that $f(Q) \cap Q = \emptyset$ (indeed, $f(\overline{\partial D \setminus X}) \cap D = \emptyset$ implies that $f(Q) \cap Q = \emptyset$). Thus, $\text{var}(f,Q)$ is well-defined.

DEFINITION 7.4.5. Let f be a positively oriented map and X a continuum. If $f(p) = p$ and $p \in \partial X$ then we say that f *repels outside* X *at* p provided there exists a ray $R \subset \mathbb{C} \setminus X$ from ∞ which lands on p and a sequence of simple closed curves S^j bounding closed disks D^j such that $D^1 \supset D^2 \supset \ldots$, $p \in \text{int}(D^j)$, $\cap D^j = \{p\}$, $f(D^1 \cap X) \subset X$, $f(\overline{S^j \setminus X}) \cap D^j = \emptyset$ and for each j there exists a component Q^j of $S^j \setminus X$ such that $Q^j \cap R \neq \emptyset$ and $\text{var}(f, Q^j) \neq 0$.

Definition 7.4.5 gives some information about dynamics around p.

LEMMA 7.4.6. *Suppose that $f : \mathbb{C} \to \mathbb{C}$ is a positively oriented map, X a non-separating continuum, $p \in \partial X$ such that $f(p) = p$ and f repels outside X at p. If R is the ray from the Definition 7.4.5 then $f(Q^j) \cap R \neq 0$. Moreover, if f scrambles the boundary of X, then $\text{var}(f, Q^j) > 0$.*

Thus, even though R above may be non-invariant, there are crosscuts approaching p which are mapped by f "along R farther away from p".

PROOF. Take $z \in R \cap Q^j$ so that $(z, \infty)_R \cap Q^j = \emptyset$. Choose a junction with $[z, \infty]_R$ (the subray of R running from z to infinity) playing the role of R_i and two other rays close to $[z, \infty]_R$ on both sides. Then $f(Q^j) \cap R = \emptyset$ implies $\mathrm{var}(f, Q^j) = 0$, a contradiction. The second claim follows by Lemma 7.3.2. □

We will use the following version of uniformization. Let X be a non-separating continuum and $\varphi : \mathbb{D}_\infty \to \mathbb{C} \setminus X$ an onto conformal map such that $\varphi(\infty) = \infty$ (here $\mathbb{D}_\infty = \mathbb{C} \setminus \overline{\mathbb{D}}$ is the complement of the closed unit disk). Thus, we choose the uniformization, under which the *complement* $\mathbb{C} \setminus \overline{\mathbb{D}}$ of the closed unit disk corresponds to the *complement* $\mathbb{C} \setminus X$ of X. Of course, the same can be considered on the two-dimensional sphere $\mathbb{C}^\infty$ which is sometimes more convenient. Notice, that since $\overline{\mathbb{D}}$ is a non-separating continuum in $\mathbb{C}$, we can use for it the usual terminology (crosscuts, shadows, etc). Also recall that the shadow of a crosscut C of a nonseparating continuum X is the bounded component of $\mathbb{C} \setminus [X \cup C]$ (and *not* its closure).

Then, given a crosscut C of X with endpoints x, y, we can associate to its endpoints external angles as follows. It is well known [**Mil00**] that $\varphi^{-1}(C)$ is a crosscut of the closed unit disk with endpoints α, β. It follows that we can extend φ by defining $\varphi(\alpha) = x$ and $\varphi(\beta) = y$. Note that this extension is not necessarily continuous. In this case we say that α *corresponds* to x and β *corresponds* to y. There is a unique arc $I \subset \mathbb{S}^1$ with endpoints α, β, contained in the shadow of $\varphi^{-1}(C)$. Assuming that the positive orientation on I is from α to β, we choose the appropriate orientation of C (i.e., in this case from x to y) and call such an oriented C *positively oriented*.

Observe that in this situation if D is a disk around a point $x \in X$ then components of $\partial D \setminus X$ are crosscuts of X whose φ-preimages are crosscuts of $\overline{\mathbb{D}}$ in the uniformization plane. However, these preimage-crosscuts in $\mathbb{D}_\infty$ may be located all over $\mathbb{S}^1$.

The next theorem is the main result of this subsection.

THEOREM 7.4.7. *Suppose that f is a positively oriented map of the plane with only isolated fixed points, $X \subset \mathbb{C}$ is a non-separating continuum or a point, and the following conditions hold.*

(1h) *For each fixed point $p \in X$ we have that $x \in \partial X$, $\mathrm{ind}(f, p) = 1$ and f repels outside X at p.*

(2h) *The map f scrambles the boundary of X. Moreover, using the notation from Definition 5.4.1 it can be said that for each i either $f(K_i) \cap Z_i = \emptyset$, or there exists a neighborhood U_i of K_i with $f(U_i \cap X) \subset X$.*

Then X is a point.

PROOF. Suppose that X is not a point. Since f has isolated fixed points, there exists a simply connected neighborhood V of X such that all fixed points $\{p_1, \ldots, p_m\}$ of $f|_{\overline{V}}$ belong to X. The idea of the proof is to construct a tight bumping simple closed curve S such that $X \subset T(S) \subset V$ and $\mathrm{var}(f, S) \geq m$. Hence $\mathrm{ind}(f, S) = \mathrm{var}(f, S) + 1 \geq m + 1$ while by Theorem 7.4.4 our assumptions imply that $\mathrm{ind}(f, S) = m$, a contradiction.

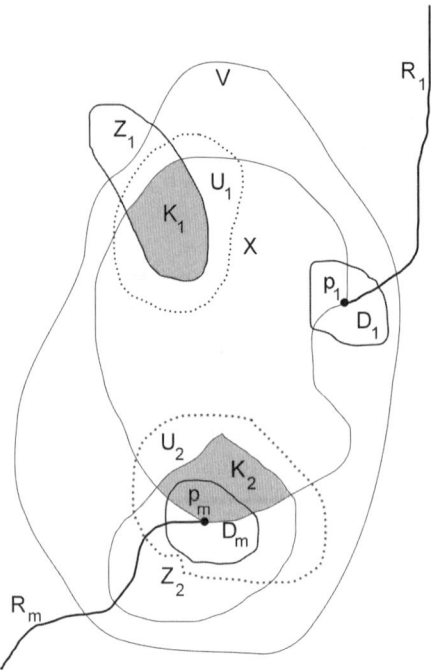

FIGURE 7.2. Illustration to the proof of Theorem 7.4.7.

First we need to make a few choices of neighborhoods and constants; we assume that there are n exit continua $K_1, \ldots, K_n$. We also assume that they are numbered so that for $1 \leq i \leq n_1$ we have $f(K_i) \cap Z_i = \emptyset$ and for $n_1 < i \leq n$ we have $f(K_i) \cap Z_i \neq \emptyset$. Choose for each i a small neighborhood U_i of K_i as follows.

CHOOSING NEIGHBORHOODS U_i OF EXIT CONTINUA K_i

(1) By assumption (2) of the theorem we may assume that $f(U_i \cap X) \subset X$ for each i with $n_1 < i \leq n$.
(2) By continuity we may assume that $d(U_i \cup Z_i, f(U_i)) > 0$ for $1 \leq i \leq n_1$.
(3) We may assume that $T(X \cup \bigcup U_i) \subset V$ and $\overline{U_i} \cap \overline{U_k} = \emptyset$ for all $i \neq k$.
(4) We may assume that every fixed point of f contained in $\overline{U_i}$ is contained in K_i.

Let $\{p_1, \ldots, p_t\}$ be all fixed points of f in $X \setminus \bigcup_i K_i$ and let $\{p_{t+1}, \ldots, p_m\}$ be all the fixed points contained in $\bigcup K_i$. Observe that then by part (4) of the choice of neighborhoods U_i we have $p_i \in X \setminus \overline{\bigcup U_s}$ if $1 \leq i \leq t$. Also, it follows that for each $j, t+1 \leq j \leq m$, there exists a unique $r_j, 1 \leq r_j \leq n$, such that $p_j \in K_{r_j}$. For each fixed point $p_j \in X$ we rely upon Definition 7.4.5 and, as specified in that definition, choose a ray $R_j \subset \mathbb{C} \setminus X$ landing on p_j. Now we choose closed disks D_j around each p_j from Definition 7.4.5 so that, in addition to properties from Definition 7.4.5 (listed below as (6), (9) and (10)), they satisfy the following conditions.

THE CHOICE OF CLOSED DISKS D_j

(5) $D_j \cap R_i = \emptyset$ for all $i \neq j$.
(6) $f(\overline{S_j \setminus X}) \cap D_j = \emptyset$.

(7) $T(X \cup \bigcup_j D_j) \subset V$.
(8) $[D_j \cup f(D_j)] \cap [D_k \cup f(D_k)] = \emptyset$ for all $j \neq k$.
(9) $f(D_j \cap X) \subset X$ (this is possible because f repels outside X at each fixed point of f and by Definition 7.4.5).
(10) Denote by $Q(j,s)$ all components of $S_j \setminus X$; then there exists a component, $Q(j,s(j))$, of $S_j \setminus X$, with $\text{var}(f, Q(j,s(j))) > 0$ and $Q(j,s(j)) \cap R_j \neq \emptyset$ (this is possible by Definition 7.4.5 and Lemma 7.4.6).
(11) $[D_j \cup f(D_j)] \cap \bigcup \overline{U_i} = \emptyset$ for all $1 \leq j \leq t$.
(12) If $t < j \leq m$ then $[D_j \cup f(D_j)] \subset U_{r_j}$.

Let us use our standard uniformization $\varphi : \mathbb{D}_\infty \to \mathbb{C} \setminus X$ described before the statement of the theorem. It serves as an important tool; in particular it allows us to pull crosscuts from the plane containing X to $\mathbb{D}_\infty$ and introduce the appropriate orientation on all these crosscuts.

Claim A. *Suppose that $W \subset D_j$ is a Jordan disk around p_j (e.g., W may coincide with D_j) and C is a crosscut which is a component of $\partial W \setminus X$. Then the shadow of $Q(i,s(i)), i \neq j$ is not contained in the shadow of C (thus, the shadows of $Q(j,s(j))$ and $Q(i,s(i))$) are disjoint). Moreover, if W is sufficiently small, then the shadow of $Q(j,s(j))$ is not contained in the shadow of C either.*

Proof of Claim A. By condition (8) from the choice of the disks D_j those disks are pairwise disjoint. Hence all the crosscuts $Q(r,t)$ are pairwise disjoint, and $C \cap Q(i,s(i)) = \emptyset$. If the shadow of $Q(i,s(i))$ is contained in the shadow of C, then the ray R_i intersects C, contradicting condition (5) from the choice of the disks D_j. Hence $\text{Sh}(Q(j,s(j))) \cap \text{Sh}(Q(i,s(i))) = \emptyset$ for $i \neq j$ (otherwise, because the crosscuts are pairwise disjoint, one of the shadows would contain the other one, impossible by the just proven). Now, if $\text{Sh}(Q(j,s(j))) \subset \text{Sh}(C)$, then, in $\mathbb{D}_\infty$, $\varphi^{-1}(C)$ shields $\varphi(Q(j,s(j)))$ from infinity and must be, together with C, of a bounded away from zero size. Hence, if W is very small, this cannot happen. □

Now we define another collection of disks around the points p_j. By Claim A for each j we choose a small Jordan disk D'_j from Definition 7.4.5 around p_j so that no shadow $\text{Sh}(Q(i,s(i))$ is contained in the shadow of any crosscut C which is a component of $(\partial D'_j) \setminus X$. In particular, for each such C, $\overline{f(C)} \cap \overline{C} = \emptyset$. Let us now choose a few constants.

THE CHOICE OF CONSTANTS $\eta < \delta < \varepsilon$

(13) Choose $\varepsilon > 0$ such that for all $x \in X \setminus \bigcup D'_j$, $d(x, f(x)) > 3\varepsilon$ and for each crosscut C of X of diameter less than ε with at least one endpoint outside of $\bigcup D_j$ we have that $f(C)$ is disjoint from C (observe that outside any given neighborhood of $\{p_1, \ldots, p_m\}$ all points of X move under f by a bounded away from zero distance).
(14) Choose $\delta > 0$ so that the following several inequalities hold:
  (a) $3\delta < \varepsilon$,
  (b) $3\delta < d(Z_i, Z_j)$ for all $i \neq j$,
  (c) $3\delta < d(Z_i, [X \cup f(X)] \setminus [Z_i \cup U_i])$ for each i,
  (d) $3\delta < d(K_i, \mathbb{C} \setminus U_i)$ for each i,
  (e) if $f(K_i) \cap Z_i = \emptyset$, then $3\delta < d(f(U_i), Z_i \cup U_i)$.
(15) By continuity choose $\eta > 0$ such that for each set $H \subset V$ of diameter less than η we have $\text{diam}(H) + \text{diam}(f(H)) < \delta$ and that $d(D'_i, D'_j) > \eta, i \neq j$.

By (11) and (13) above, if a set $H \subset V$ is of diameter at most η and $H \not\subset \bigcup D'_i$, then $f(H) \cap H = \emptyset$. Indeed, otherwise let $x \in H \setminus \bigcup D'_i$ and $y \in H$ be such that $f(y) \in H$. Then by (13) $d(x, f(x)) > 3\varepsilon$ while by (15) and the triangle inequality $d(x, f(x)) \le d(x, f(y)) + d(f(y), f(x)) < \delta < \varepsilon/3$, a contradiction.

Consider the family $\mathcal{E}_X$ of all components of the sets $(\partial D'_i) \setminus X$, and the crosscuts $Q(i, s(i))$. By condition (8) from the choice of the closed disks D_j, the disks D_j are pairwise disjoint; hence, the crosscuts in $\mathcal{E}_X$ are pairwise disjoint.

Let T be the topological hull $T = T(X \cup (\bigcup D'_j) \cup \bigcup Q(i, s(i)))$. Then T is a non-separating continuum. Call $C \in \mathcal{E}_X$ an *unshielded (crosscut of X)* if it is a part of ∂T and denote the family of all such crosscuts by $\mathcal{E}_X^u$. By Claim A all crosscuts $Q(i, s(i))$ are unshielded. Call φ-preimages of unshielded crosscuts *unshielded (crosscuts of $\overline{\mathbb{D}}$)* and denote their family by $\mathcal{E}_\mathbb{D}^u$. Clearly, any two unshielded crosscuts have disjoint shadows.

For each $C \in \mathcal{E}_X^u$, let $C_\mathbb{D} = \varphi^{-1}(C)$. Note that there are at most finitely many crosscuts $C \in \mathcal{E}_X^u$ with $\text{diam}(C) \ge \eta/30$. Let $C^1, \ldots, C^q$ be the collection of all crosscuts $Q(i, s(i))$ and all crosscuts in $\mathcal{E}_X^u$ with diameters at least $\eta/30$. By definition 7.4.5, $f(\overline{C^j}) \cap \overline{C^j} = \emptyset$ for each j. Then the crosscuts $C_\mathbb{D}^j = \varphi^{-1}(C^j)$ are all pairwise disjoint and have disjoint shadows. Hence we may assume that, if the endpoints of $C_\mathbb{D}^j$ are α_j, β_j, then $\alpha_1 < \beta_1 < \cdots < \alpha_q < \beta_q < \alpha_{q+1} = \alpha_1$ in the positive circular order around $\mathbb{S}^1$.

For each i, $1 \le i \le q$, choose a finite chain of crosscuts F_j^i in $\mathbb{D}_\infty$ with endpoints γ_j, γ_{j+1} where $\beta_i = \gamma_1 < \gamma_2 < \cdots < \gamma_k = \alpha_{i+1}$ so that all closures of crosscuts from the collection $\{C_\mathbb{D}^1, \ldots, C_\mathbb{D}^q\} \cup \{F_j^i\}_{i,j}$, $1 \le j \le k(i)$ (except for the adjacent crosscuts which share endpoints) and their shadows are pairwise disjoint (this can be easily done, e.g. because accessible points on the boundary of X are dense), $\varphi(F_j^i) = G_j^i$ is a crosscut of X and $\text{diam}(G_j^i) < \eta/30$ for all i, j. In addition we may assume that non-adjacent G_j^i have disjoint sets of endpoints. Let $Y = T(\bigcup C_\mathbb{D}^i \cup \bigcup F_j^i)$. Then Y is a Jordan disk whose boundary is a simple closed curve $\widehat{S}'' \subset \overline{\mathbb{D}_\infty}$. Let $S'' = \varphi(\widehat{S}'')$. The set $S'' \cap X$ is finite. It partitions S'' into links which include all C^i's. However, some links of the form G_j^i may be very close to a fixed point of f and may not move off themselves under f. Hence we modify S'' as follows.

Claim B. *There exists a bumping simple closed curve S such that:*

(1b) $\bigcup_{j=1}^q C^j \subset S$.

(2b) *all components of $S \setminus [X \cup \bigcup_i D'_i \cup \bigcup C^i]$ are of diameter less than η,*

(3b) *for each i components of $S \cap \text{int}(D'_i)$ are so small that they stay far away from the fixed points and are moved off themselves by f.*

Let $Z = T(S'' \cup \bigcup D'_j)$. Then Z is a Jordan disk whose boundary is a simple closed curve S', and all crosscuts C^i are still contained in S'. We modify S', keeping all C^i's but changing $S' \setminus \bigcup C^i$ so that the resulting bumping simple closed curve S can be partitioned into finitely many links each of which does not go deep into the interior of $\bigcup D'_j$ and, hence, moves off itself under f.

Consider the crosscuts $\mathcal{E}_\mathbb{D}^u$ in $\mathbb{D}_\infty$. If the chain $\{F_1^i, \ldots, F_{k(i)}^i\}$ intersects a crosscut $Q \in \mathcal{E}_\mathbb{D}^u$ let p_Q and r_Q be the first and last point of intersection of the arc $\cup_i F_j^i$ and Q. Then $p_Q \ne r_Q$. If $\varphi([p_Q, r_Q])$ is small then move forward along S''. Otherwise suppose that the endpoints of $\varphi(Q)$ are a_Q and b_Q and assume that $a_Q < b_Q$ in the positive order around X. Suppose that $p_Q \in F_j^i$ which has endpoints

γ_j and γ_{j+1} and $r_Q \in F_k^i$ which has endpoints γ_k and γ_{k+1} and $\gamma_{j+1} \leq \gamma_k$. Replace the subarc from γ_j up to γ_{k+1} in S'' by an arc joining the same endpoints whose φ-image is very close to $\varphi(Q)$. Moving forward along S'' in the positive direction and making finitely often similar modifications, we obtain the desired simple closed curve S. This completes the proof of Claim B. □

We want to compute the variation of S. Each link $Q(j, s(j))$ contributes at least 1 towards $\text{var}(f, S)$, and we want to show that all other links have non-negative variation. To do so we want to apply Lemma 7.3.2. Hence we need to verify that all links of S are bumping arcs whose endpoints map back into X such that their images are disjoint from themselves. By Claim B, all links of S move off themselves. However some links of S may have endpoints mapped off X. To ensure that for our bumping simple closed curve endpoints e of its links map back into X we have to replace some of the finite chains of links of S by one link which is their concatenation (this is similar to what was done in Theorem 7.3.3). Then we will have to check if the new "bigger" links still have images disjoint from themselves.

Denote by A the initial partition of S into links which are called A-links.

Claim C. *There exists a partition A' of S whose links are bumping arcs with endpoints mapped back into X and whose images are disjoint from themselves. Moreover, A'-links are concatenations of A-links of S such that all A-links of S of type $Q(i, t)$ remain A'-links of S.*

Proof of Claim C. Suppose that $X \cap S = A = \{a_0 < a_1 < \cdots < a_n\}$ and $a_0 \in A$ is such that $f(a_0) \in X$ (by arguments similar to those in Theorem 7.3.3 one can show that we can make this assumption without loss of generality). We only need to enlarge and concatenate links of S with at least one endpoint of the link mapped outside X. Therefore all links of S of the form $Q(j, s)$ do not come under this category of links because by Definition 7.4.5 their endpoints do map into X. Other links, however, may have endpoints mapped outside X. Observe, that by the construction all those other links are less than η in diameter and hence have the property (15) of the choice of the constants.

Let t' be minimal such that $f(a_{t'}) \notin X$ and $t'' > t'$ be minimal such that $f(a_{t''}) \in X$. Then $f(a_{t'}) \in Z_i$ for some i. Since every component of $[a_{t'}, a_{t''}] \setminus X$ has image of diameter less than δ (which is less than the distance between any two sets Z_r, Z_l), $f(a_t) \in Z_i \setminus X$ for all $t' \leq t < t''$. On the other hand, for $t' \leq t < t''$, $a_t \notin U_i$. To see this, note that if $f(K_i) \cap Z_i = \emptyset$, then by the above made choices $f(U_i) \cap Z_i = \emptyset$, and if $f(K_i) \cap Z_i \neq \emptyset$, then $f(U_i \cap X) \subset X$ by the assumption. Thus, all points $a_t, t' \leq t < t''$ are in $X \setminus U_i$ while all their images $f(a_t)$ are in $Z_i \setminus X$, which by the property (14c) of the constant δ implies that these two finite sets of points are at least 3δ distant.

Now, by the proven above, all the A-links of S in the arc $[a_{t'-1}, a_{t''}]$ are of diameters less than η. Hence it follows from the properties (15) and (14c) of the constant δ that $f([a_{t'-1}, a_{t''}]) \cap [a_{t'-1}, a_{t''}] = \emptyset$ and we can remove the points a_t, for $t' \leq t < t''$ from the partition A of S. By continuing in the same fashion we obtain a subset $A' \subset A$ such that for the closure of each component C of $S \setminus A'$, $f(C) \cap C = \emptyset$ and for both endpoints a and a' of C, $\{f(a), f(a')\} \subset X$. Moreover, as was observed above, the enlarging of links of S does not concern any links of the original bumping simple closed curve of the form $Q(j, s)$, in particular for each j, $Q(j, j(s))$ is an A'-link of S. □

Now we apply a version of the standard argument from the proof of Theorem 7.1.3; here, instead of Theorem 3.1.4 we use the fact that f satisfies the argument principle. Indeed, by Theorem 3.2.2 and Lemma 7.3.2, $\mathrm{ind}(f, S) \geq \sum_j \mathrm{var}(f, Q(j, j(s))) + 1 \geq m + 1$ contradicting Theorem 7.4.4. □

It is possible to use a different approach in Definition 7.4.5 and Theorem 7.4.7. Namely, a version of Definition 7.4.5 could define repelling outside X at a fixed point p as the existence of a family of closed disks D^j with similar properties *except* now we would require the existence of at least $\mathrm{ind}(f, p)$ pairwise non-homotopic rays outside X from ∞ to p (landing on p) and the existence of the same number of components of $S^j \setminus X$ non-disjoint from corresponding rays and with non-zero variation on each such component.

Then a version of Theorem 7.4.7 would state that if a positively oriented map with isolated fixed points f repels outside X at each of its fixed points, and the condition (2h) of Theorem 7.4.7 is satisfied, then X must be a point. The proof of this version of Theorem 7.4.7 is almost the same, except for a bit heavier notation needed (now that we have not one, but $\mathrm{ind}(f, p)$ crosscuts with positive variation around each fixed point in X. Since for our applications Theorem 7.4.7 suffices we restricted ourselves to this case.

Theorem 7.4.7 implies the following.

COROLLARY 7.4.8. *Suppose that f is a positively oriented map with isolated fixed points, and $X \subset \mathbb{C}$ is a non-separating and non-degenerate continuum satisfying condition (2h) stated in Theorem 7.4.7 and such that all fixed points in X belong to ∂X. Then either f does not repel outside X at one of its fixed points, or the local index at one of its fixed points is not equal to 1.*

Lemma 7.4.9 gives a verifiable sufficient condition for a fixed point a belonging to a locally invariant continuum X to be such that the map f repels outside X. We will apply the lemma in the next section.

LEMMA 7.4.9. *Suppose that f is a positively oriented map, $X \subset \mathbb{C}$ is a non-separating continuum or a point and $p \in \partial X$ is a fixed point of f such that:*
  (i) *there exists a neighborhood U of p such that $f|_{\overline{U}}$ is one-to-one and $f(\overline{U} \cap X) \subset X$,*
  (ii) *there exists a ray $R \subset \mathbb{C}^\infty \setminus X$ from infinity such that $\overline{R} = R \cup \{p\}$, $f|_R : R \to R$ is a homeomorphism and for each $x \in R$, $f(x)$ separates x from ∞ in R,*
  (iii) *there exists a nested sequence of closed disks $D_j \subset U$ with boundaries S_j containing p in their interiors such that $\bigcap D_j = \{p\}$ and $f(S_j \setminus X) \cap D_j = \emptyset$.*

Then for a sufficiently large j there exists a component C of $S_j \setminus X$ with $C \cap R \neq \emptyset$ and $\mathrm{var}(f, C) > 0$, so that f repels outside X at p.

Observe that here we show that $\mathrm{var}(f, C) > 0$ without any "scrambling" assumptions on f.

PROOF. Choose a Jordan disk U as in (i) so that $(\partial U) \cap R = \{q\}$ is a point and $X \setminus \overline{U} \neq \emptyset$. Choose j so that $D_j \cup f(D_j) \subset U$. By [**BO06**] there is a component C of $(\partial D_j) \setminus X$ such that R crosses C essentially (see Definition 3.6.4). Slightly adjusting D_j, we may assume that $R \cap \partial D_j$ is finite and each intersection is transversal. Since

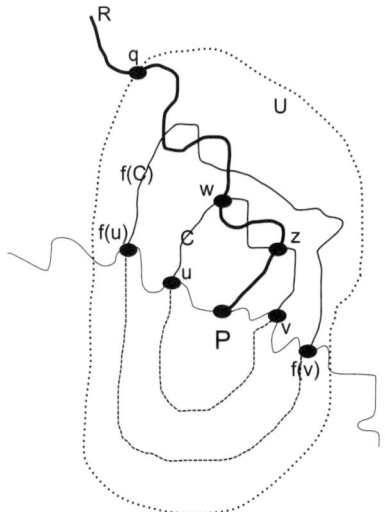

FIGURE 7.3. Illustration to the proof of Lemma 7.4.9.

R crosses C essentially, $|R \cap C|$ is odd; since f is one-to-one on $\overline{U}$, $|f(C) \cap R|$ is odd as well.

Let u, v be the endpoints of C. Observe, that C can be included in a simple closed curve S around X so that $X \subset T(S)$. Since by (i) $f(\overline{U} \cap X) \subset X$, we see that $f(u) \in T(S), f(v) \in T(S)$ and variation $\text{var}(f, C) > 0$ is well-defined (see Definition 2.2.2).

Let us move along R from infinity to p and denote by w be the first point of $C \cap R$ which we meet and by z the last point. Then by (ii), $f(C) \cap [p, z]_R = \emptyset$. Also, by (i) $|f(C) \cap R| = |f(C \cap R)|$ is odd. Since $X \setminus D_j \neq \emptyset$ and $X \cap ([z, w]_C \cup [z, w]_R) = \emptyset$, X is contained in the unbounded component V^∞ of the complement to $[z, w]_C \cup [z, w]_R$. Since u and v belong to $U \cap X$, their images $f(u)$ and $f(v)$ belong to X as well. Hence $f(u), f(v) \in V^\infty$.

Claim A. $|f(C) \cap [z, w]_R|$ *is even.*

Proof of Claim A. A complementary domain O of $[z, w]_C \cup [z, w]_R$ is called *even/odd* if there is an arc J from infinity to a point in O so that $J \cap C \cap R = \emptyset$, $J \cap ([z, w]_C \cup [z, w]_R)$ is finite, every intersection is transversal and $|J \cap ([z, w]_C \cup [z, w]_R)|$ is even/odd, respectively. By [**OT82**] the notion of an even/odd domain is independent of J, well-defined and each complementary domain of $[z, w]_C \cup [z, w]_R$ is either even or odd. Since by (iii) $f(\overline{C}) \cap \overline{C} = \emptyset$, $f(\overline{C})$ can only intersect $C \cup R$ at points of R. Also, whenever $f(\overline{C})$ meets $[z, w]_R$, it crosses from an even to an odd domain or vice versa. Since both $f(u)$ and $f(v)$ are in the unbounded (and hence even) domain of $\mathbb{C} \setminus [f(C) \cup [z, w]_R]$, $|f(C) \cap [z, w]_R|$ is even as desired. □

Observe that $f(\overline{C})$ is outside D_j and hence is disjoint from $[p, z]_R \subset D_j$. Since $|f(C) \cap R|$ is odd and $|f(C) \cap [p, w]_R| = |f(C) \cap [z, w]_R|$ is even, $|f(C) \cap [w, \infty]_R|$ is odd. Since every intersection is transversal, we can replace $[w, \infty]_R$ by a junction

J_w such that every intersection point of $f(C) \cap [w, \infty)_R$ contributes exactly $+1$ or -1 to the count. Hence $\mathrm{var}(f, C) \neq 0$.

We prove next that $\mathrm{var}(f, C) > 0$. Note that $[p, q]_R \subset \overline{U}$ is an arc which meets ∂U only at q. Let W be the component of $U \setminus (C \cup R)$ whose boundary contains $(\partial U) \setminus q$. Note that u and v are accessible points in ∂W. Hence points u and v can be connected with an arc K inside W disjoint from $R \cup C$. Then, since $f|_U$ is a homeomorphism, $f(K) \cap R = \emptyset$. By Lemma 7.3.1 this implies that $\mathrm{var}(f, C) \geq 0$ and by the previous paragraph then $\mathrm{var}(f, C) > 0$ (basically, we simply choose a junction J'_w close to $[w, \infty)_R$ such that $f(K) \cap J'_w = \emptyset$ and conclude that $\mathrm{var}(f, C) = \mathrm{win}(f, C \cup K, w) > 0$ since f is a positively oriented map). □

It is now easy to see that the following corollary holds.

COROLLARY 7.4.10. *Suppose that $X \subset \mathbb{C}$ is a non-separating continuum or a point and $f : \mathbb{C} \to \mathbb{C}$ is a positively oriented map with isolated fixed points, and the following conditions hold.*

(a) *Each fixed point $p \in X$ is topologically repelling, belongs to ∂X, and has a neighborhood U_p such that $f(U_p \cap X) \subset X$ and $f|_{U_p}$ is a homeomorphism.*
(b) *For each fixed point $p \in X$, there exists a ray $R \subset \mathbb{C}^\infty \setminus X$ from infinity landing on p, $f|_R : R \to R$ is a homeomorphism and for each $x \in R$, $f(x)$ separates x from ∞ in R.*
(c) *The map f scrambles the boundary of X. Moreover for every i either $f(K_i) \cap Z_i = \emptyset$ or there exists a neighborhood U_i of K_i with $f(U_i \cap X) \subset X$.*

Then X is a (fixed) point.

PROOF. Let us apply Theorem 7.4.7. To do so, we verify its conditions. The facts that $X \subset \mathbb{C}$ is a non-separating continuum or a point and $f : \mathbb{C} \to \mathbb{C}$ is a positively oriented map with isolated fixed points are clearly satisfied. To verify condition (1h) of Theorem 7.4.7, suppose that $p \in X$ is a fixed point. Then by (a) above $p \in \partial X$. Moreover, p is topologically repelling, and so by Lemma 7.4.3 the index at p is $+1$.

It remains to verify that f repels outside X at p. To do so we apply Lemma 7.4.9. Since p is topologically repelling, there exists a nested sequence of closed disks $D_j \subset U$ with boundaries S_j containing p in their interiors, with $\bigcap D_j = \{p\}$ and $\overline{f(S^j \setminus X)} \cap D^j = \emptyset$. Hence the condition (iii) of Lemma 7.4.9 is satisfied. The condition (i) of Lemma 7.4.9 immediately follows from (a) above; the condition (ii) of Lemma 7.4.9 immediately follows from (b) above. Hence by Lemma 7.4.9 the map f repels outside X at p. Therefore the condition (1h) of Theorem 7.4.7 is satisfied. Condition (2h) of Theorem 7.4.7 is also satisfied (it simply coincides with condition (c) of our corollary), hence by Theorem 7.4.7 X is a point. □

7.5. Applications to complex dynamics

We begin by introducing a few facts concerning local dynamics at parabolic and repelling periodic points of a polynomial which were not necessary for stating the results of this section in Chapter 5 but are needed for the proofs. A nice description of this can be found in [**Mil00**] ([**CG93**] can also serve as a good source here).

Let $P : \mathbb{C} \to \mathbb{C}$ be a complex polynomial, J_P its Julia set (J_P is the closure of the set of repelling periodic points of P) and $K_P = T(J_P)$ the "filled-in" Julia set. Recall That $\sigma_d : \mathbb{S}^1 \to \mathbb{S}^1$ is defined by $\sigma_d(\alpha) = d\alpha \mod 1$, where $\mathbb{S}^1 = \mathbb{R}/\mathbb{Z}$

is parameterized by $[0,1)$. (This map is conjugate to the map $z \to z^d$ restricted to the unit circle in the complex plane.) If p is a periodic point of P of period n and $(P^n)'(p) = re^{2\pi i\alpha}$ with $r \geq 0$, then p is *repelling* if $r > 1$, *parabolic* if $r = 1$ and $\alpha \in \mathbb{Q}$, *irrational neutral* if $r = 1$ and $\alpha \in \mathbb{R} \setminus \mathbb{Q}$ and *attracting* if $r < 1$. If p is a repelling or parabolic fixed point in a non-degenerate component Y of K_P, then by [**DH85a, LP96**] there exist $1 \leq k < \infty$ external rays $R_{\alpha(i)}$ such that $\sigma_d|_{\{\alpha(1),\ldots,\alpha(k)\}} : \{\alpha(1),\ldots,\alpha(k)\} \to \{\alpha(1),\ldots,\alpha(k)\}$ is a permutation, all $\alpha(i)$ are of the same minimal period under σ_d, for each j the ray $R_{\alpha(j)}$ lands on p, and no other external rays land on p.

Components of $\mathbb{C} \setminus J_P$ are called *Fatou domains*. There are three types of bounded Fatou domains U. A Fatou domain U is called an *attracting domain* if it contains an attracting periodic point, a *Siegel domain* if it contains an irrational neutral periodic point and a *parabolic domain* if it is periodic but contains no periodic points. In the latter case there always exists a parabolic periodic point on the boundary of the parabolic Fatou domain. An irrational neutral periodic point inside a Siegel domain is called a *Siegel (periodic) point*; an irrational neutral periodic point in J_P is called a *Cremer (periodic) point*.

Any two distinct parabolic Fatou domains which contain the same parabolic periodic point in their boundaries are separated by two external rays which land at this parabolic point. It is also known that points inside these parabolic domains are attracted by the orbit of this parabolic periodic point while points on the external rays landing at points of this orbit are repelled to infinity. Suppose that p is a parabolic fixed point, $P'(p) = e^{2\pi i \frac{r}{q}}$, $r, q \in \mathbb{Z}$, $R_0, \ldots, R_{m-1}$ are all external rays landing at p and $U_0, \ldots, U_{k-1}$ are all Fatou domains which contain p in their boundaries. Moreover, suppose that both rays and domains are numbered according to the positive circular order around p. Then combinatorially one can think of the local action of P on rays and domains at p as a rotation by r/q. This means that $P(U_j) = U_{(j+r) \mod k}$ and all rays between U_i and U_{i+1} are mapped by P in an order preserving way onto all rays between $U_{(i+r) \mod k}$ and $U_{(i+1+r) \mod k}$.

Before we continue we want to recall the notion of a general puzzle-piece which is first introduced in Definition 5.5.1.

DEFINITION 5.5.1 (General puzzle-piece). Let $P : \mathbb{C} \to \mathbb{C}$ be a polynomial. Let $X \subset K_P$ be a non-separating subcontinuum or a point such that the following holds.

(1) There exists $m \geq 0$ and m pairwise disjoint non-separating continua/points $E_1 \subset X, \ldots, E_m \subset X$.
(2) There exist m finite sets of external rays $A_1 = \{R_{a_1^1}, \ldots, R_{a_{i_1}^1}\}, \ldots, A_m = \{R_{a_1^m}, \ldots, R_{a_{i_m}^m}\}$ with $i_k \geq 2, 1 \leq k \leq m$.
(3) We have $\Pi(A_j) \subset E_j$ (so the set $E_j \cup (\cup_{k=1}^{i_j} R_{a_k^j}) = E'_j$ is closed and connected).
(4) X intersects a unique component C_X of $\mathbb{C} \setminus \cup E'_j$.
(5) For each Fatou domain U either $U \cap X = \emptyset$ or $U \subset X$.

We call such X with the continua E_i and the external rays $R_{\alpha_i^k}$ a *general puzzle-piece* and call the continua E_i *exit continua* of X. For each k, the set E'_k divides the plane into i_k open sets which we will call *wedges* (at E_k); denote by W_k the wedge which contains $X \setminus E_k$ (it is well-defined by (4) above).

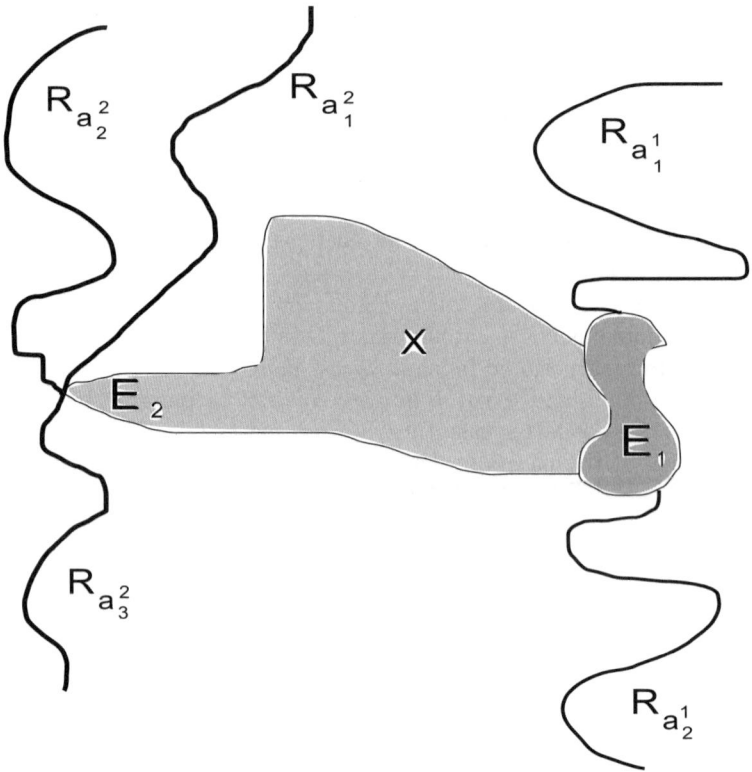

FIGURE 7.4. A general puzzle-piece

Let us now see whether condition (1) of Theorem 7.4.7 applies to the polynomial P with fixed parabolic points in a general puzzle-piece X such that $P(X) \cap C_X \subset X$ (loosely, it means that $P(X)$ does not "grow" at points of C_X). We need to check that at a fixed parabolic point $p \in X$ the map P repels outside X and the local index is 1.

For convenience, consider $p \in C_X$. Let L be a cycle of parabolic domains containing p in their boundaries. To make a more complete picture, we first observe that either $L \subset X$ or all domains in L are disjoint from X. Indeed, suppose that one of the domains in L is contained in X. Then, as there is no "growth" of $P(X)$ at p, we see that all domains of L are in X. Otherwise by condition (5) from Definition 5.5.1 all domains in L are disjoint from X.

Suppose first that $P'(p) \neq 1$. There must exist small disks D such that components of $(\partial D) \setminus X$ map outside D. Let Q be a crosscut which is a component of $(\partial D) \setminus X$. Choose an external ray R landing at p and crossing Q essentially. By the analysis of dynamics around p (the fact that P locally "rotates", see, e.g., [**Mil00**]) and since $P(X) \cap C_X \subset X$, we conclude that $P(Q) \cap R = \emptyset$. However, then $\text{var}(f, Q) = 0$, a contradiction with the existence of a crosscut of positive variation in $(\partial D) \setminus X$.

On the other hand, if $P'(p) = 1$ then, as was explained right after the proof of Lemma 7.4.3, $\text{ind}(P, p) > 1$ (again, see, e.g., [**Mil00**]) which contradicts the condition (1) of Theorem 7.4.7 as well. Hence in the parabolic case Theorem 7.4.7 cannot be applied "as is" to the polynomial P. Moreover, since at parabolic points the map is not topologically repelling, neither is Corollary 7.4.10 applicable in this case.

The idea allowing us to still deal with parabolic points is that we can change P inside the parabolic domains in question without compromising the rest of the arguments and modifying these parabolic points to topologically repelling periodic points. The thus constructed new map g will no longer be holomorphic but will satisfy the conditions of Corollary 7.4.10. We formalize this idea in the following lemma.

LEMMA 7.5.1. *Suppose that $\{p_j\}, 1 \le j \le m$, are all parabolic fixed points of a polynomial P with $P'(p_j) = 1$. Then there exists a positively oriented map $g_P = g$ which coincides with P outside the invariant parabolic Fatou domains, is locally one-to-one at each p_j (hence p_j is not a critical point of g) and is such that all the points p_j are topologically repelling fixed points of g. In particular, $\text{ind}(g, p_j) = +1$ for all j, $1 \le j \le m$.*

PROOF. Let us consider a fixed parabolic point $p = p_j$. Let F_i be the invariant Fatou domains containing p in their boundaries B_i. By a nice recent result of Yin and Roesch [**RY08**], the boundary B_i of each F_i is a simple closed curve and $P|_{B_i}$ is conjugate to the map $z \to z^{d(i)}$ for some integer $d(i) \ge 2$. Let $\psi : F_i \to \mathbb{D}$ be a conformal isomorphism. Since B_i is a simple closed curve, ψ extends to a homeomorphism on $\overline{\mathbb{D}}$. Since $f|_{B_i}$ is conjugate to the map $z \to z^{d(i)}$, it now follows that the map $P|_{\overline{F_i}}$ can be replaced by a map topologically conjugate by ψ to the map $g_i(z) = z^{d(i)}$ on the closed unit disk $\overline{\mathbb{D}}$ which agrees with f on B_i. Let g be the map defined by $g(z) = P(z)$ for each $z \in \mathbb{C} \setminus \bigcup F_i$ and $g(z) = g_i(z)$ when $z \in F_i$. Then g is clearly a positively oriented map.

Since by the analysis of the dynamics around parabolic points [**Mil00**] P repels points away from p outside parabolic domains F_i, we conclude, by the construction, that p is a topologically repelling fixed point of g. Clearly, $\deg_g(p) = 1$, hence by Lemma 7.4.3 $\text{ind}(g, p) = \deg_g(p) = 1$ as desired. Continuing in this fashion, we can change P on all invariant parabolic domains with fixed points p_j, $1 \le j \le m$, in their boundaries to a map g which satisfies the requirements of the lemma. $\square$

We use Lemma 7.5.1 in the proof of the Theorem 7.5.2. Recall, that a fixed point x of a polynomial P is said to be *non-rotational* if there is a fixed external ray landing at x (it follows that each such point is either repelling or parabolic).

THEOREM 7.5.2. *Let P be a polynomial with filled-in Julia set K_P and let Y be a non-degenerate periodic component of K_P such that $P^p(Y) = Y$. Suppose that $X \subset Y$ is a non-degenerate general puzzle-piece with $m \ge 0$ exit continua $E_1, \ldots, E_m$ such that $P^p(X) \cap C_X \subset X$ and either $P^p(E_i) \subset W_i$, or E_i is a P^p-fixed point. Then at least one of the following claims holds:*

(1) *X contains a P^p-invariant parabolic domain,*
(2) *X contains a P^p-fixed point which is neither repelling nor parabolic, or*
(3) *X has an external ray R landing at a repelling or parabolic P^p-fixed point such that $P^p(R) \cap R = \emptyset$ (i.e., P^p locally rotates at some parabolic or repelling P^p-fixed point).*

Equivalently, suppose that Y is a non-degenerate periodic component of K_P such that $P^p(Y) = Y$, $X \subset Y$ is a general puzzle-piece with $m \geq 0$ exit continua $E_1, \ldots, E_m$ such that $P^p(X) \cap C_X \subset X$ and either $P^p(E_i) \subset W_i$, or E_i is a P^p-fixed point; if, moreover, X contains only non-rotational P^p-fixed points and does not contain P^p-invariant parabolic domains, then it is degenerate.

PROOF. We may assume that $p = 1$ and $P(Y) = Y$. We show that if none of the conclusions (1)-(3) hold, then Corollary 7.4.10 applies and therefore X must be a point, contradicting the assumption that X is non-degenerate. However, if X contains parabolic points, we first use Lemma 7.5.1 and replace P by the positively oriented map g constructed in that lemma. If conclusion (1) does not hold, then X contains no invariant parabolic domains, and so $P|_X = g|_X$. Let us now check conditions of Corollary 7.4.10.

First we check condition (a) of Corollary 7.4.10. Indeed, consider a fixed point $y \in X$. Then y is a repelling or parabolic fixed point of P (this is because we assume that claim (2) of Theorem 7.5.2 does not hold and hence all P^p-fixed points in X are repelling or parabolic). By Lemma 7.5.1 this implies that y is a topologically repelling fixed point of g. Let us show that there exists a small neighborhood U of y such that $g(U \cap X) \subset X$. This is clear if $y \in C_X$ because $g(X) \cap C_X \subset X$ by our assumptions. Assume now that $y \notin C_X$ which means that $\{y\} = E_k$ is one of the exit-continua of X. Observe that there a few *fixed* external rays of P landing at y (the rays are fixed because we assume that the conclusion (3) of Theorem 7.5.2 does not hold and hence all rays which land at y must be fixed). Choose the two rays R_1, R_2 which land at y and form the boundary of the wedge W_k at y which contains X. Since $g(X) \cap C_X \subset X$ by our assumptions, this implies that in a small neighborhood U_k of E_k the intersection $U_k \cap X$ maps (by g or P) into X as desired. This completes the verification of condition (a) of Corollary 7.4.10.

Condition (b) of Corollary 7.4.10 (i.e., the existence of a fixed external ray landing at each fixed point in X) follows immediately from the assumption that claim (3) of Theorem 7.5.2 above does not hold.

Let us now check condition (c) of Corollary 7.4.10. Set $g(X) \setminus X = P(X) \setminus X = H$. We may assume that $H \neq \emptyset$ and we can think of $g(X) = P(X)$ as a continuum which "grows" out of X. In particular, $m \geq 1$. Fix k, $1 \leq k \leq m$. Since $g(X) \cap C_X \subset X$, any component of H whose closure intersects E_k must be contained in one of the wedges at E_k (such wedges are defined in Definition 7.4.10), but not in W_k. Let Z_k be the topological hull of the union of all components of H which meet E_k together with E_k. Then Z_k is a non-separating continuum. Since either $g(E_i) \subset W_i$ or E_i is a fixed point, the map g scrambles the boundary of X (see Definition 5.4.1). Moreover, if E_k is mapped into W_k then clearly $g(E_k) \cap Z_k = \emptyset$ (because $Z_k \setminus E_k$ is contained in the other wedges at E_k and is disjoint from W_k). On the other hand, consider a fixed point $y \in X$ such that $E_k = \{y\}$. Then condition (c) of Corollary 7.4.10 follows from (a) which has already been verified. Hence, by Corollary 7.4.10 we conclude, that X is a point, a contradiction. □

Notice that if X is a general puzzle-piece with $m = 0$ in Theorem 7.5.2, then $C_X = \mathbb{C}$. Hence in this case $P(X) \cap C_X \subset X$ implies $P(X) \subset X$ and X is invariant. Thus, a non-separating invariant continuum $X \subset K_P$ is a general puzzle-piece if and only if for every Fatou domain U of P either $U \cap X = \emptyset$, or $U \subset X$.

The proof of the next corollary is left to the reader.

COROLLARY 7.5.3. Suppose that Y is a non-degenerate periodic component of K_P with $P^p(Y) = Y$, $X \subset Y$ is a general puzzle-piece which is either invariant or has $m > 0$ exit continua $E_1, \ldots, E_m$ such that $X \cap C_X = K_P \cap C_X$, and either $P^{pn}(E_i) \subset W_i$ for all $n > 0$, or E_i is a periodic point. If, moreover, X contains no periodic parabolic domains, no attracting, Cremer, or Siegel periodic points, and at most finitely many periodic points with more than one external ray landing at them, then X is a point.

Theorem 7.5.2 is useful in proving the degeneracy of certain impressions and establishing local connectedness of the Julia set at some points (see [**BCO08**]). Recall that the *impression* $\text{Imp}(\alpha)$ of an angle in the connected case can be defined as the intersection of the closures of all shadows $\overline{\text{Sh}(C)}$ of all crosscuts C such that R_α crosses C essentially. Corollary 7.5.4 is proved for a somewhat larger class of subcontinua of J_P which includes impressions as an important particular case.

Consider a repelling or parabolic periodic or preperiodic point x and all external rays landing at x. Then the union of two such rays and x is said to be a *legal cut* of the plane. Also, suppose that $x, y \in \partial U$ are two periodic or preperiodic points in the boundary of an attracting or parabolic Fatou domain U. By [**RY08**] there exists an arc $A \subset U$ connecting x and y. The union of A and two external rays landing at x and y is also a *legal cut* of the plane. Finally, call a continuum Q *periodic* if for some $n > 0$ we have $P^n(Q) \subset Q$.

COROLLARY 7.5.4. Let $P : \mathbb{C} \to \mathbb{C}$ be a complex polynomial and $Q \subset J_P$ be a periodic continuum such that for every legal cut C the set $Q \setminus C$ is contained in one component of $\mathbb{C} \setminus C$. Suppose that $T(Q)$ contains no Siegel or Cremer points. Then Q is degenerate. In particular, if J_P is connected and Q is a periodic impression such that $T(Q)$ contains no Cremer or Siegel points, then Q is a point.

PROOF. By considering an appropriate power of P we may assume that Q is invariant and non-degenerate. Clearly this implies that $T(Q)$ is invariant too. Suppose that $p' \in Q$ is a fixed point of P and R_β is an external ray landing at p'. Then $P(R_\beta)$ also lands on p'. If R_β is not fixed, then $C = R_\beta \cup P(R_\beta)$ is a legal cut. The local dynamics at p' and the fact that Q is invariant imply now that Q has points on either side of C, a contradiction with the assumptions on Q. Hence each fixed repelling or parabolic point in Q is non-rotational.

Let us show that Q can only intersect the closure of a parabolic or attracting Fatou domain U at one point. Indeed, suppose otherwise and let $x, y \in \partial U \cap Q, x \neq y$. Then there exists an arc $I \subset \partial U$ with endpoints x, y, contained in Q because otherwise Q will "shield" some points of ∂U from infinity contradicting the fact that all points of ∂U belong to the closure of the basin of attraction of infinity. By [**RY08**] we can find, say, periodic points $u, v \in I$ and include them in a legal cut T which will separate some points of I (and hence of Q) from other points of I, contradiction with our assumptions. Hence $T(Q)$ cannot contain an attracting or parabolic domain U since otherwise, by the above, Q must shield part of ∂U from the basin of attraction of infinity, a contradiction. This implies that $T(Q)$ cannot contain parabolic domains or attracting points. By the assumption $T(Q)$ does not contain Cremer or Siegel points either. Hence by Corollary 7.5.3 Q is a point as desired. □

In the particular case in the end of the statement we assume that J_P is connected; the same result in fact holds for *all* Julia sets but will require the introduction of the notion of the impression for disconnected Julia sets which we avoid here for the sake of simplicity. The verification of the fact that impressions satisfy the conditions of the corollary is straightforward and therefore is left to the reader. In particular, suppose R_α is a periodic external ray and the topological hull $T(\mathrm{Imp}(\alpha))$ of the impression of α contains only repelling or parabolic periodic points. Then, by Corollary 7.5.4, $\mathrm{Imp}(\alpha)$ is degenerate.

Note that the assumptions of Corollary 7.5.4 are equivalent to the following. Suppose that $Q \subset J_P$ is a periodic continuum such that for every legal cut C the set $Q \setminus C$ is contained in one component of $\mathbb{C} \setminus C$. As in the proof of Corollary 7.5.4, this implies that Q can only intersect the boundaries of attracting or parabolic domains at no more than one point. To make the conclusion of the corollary, we need to check that $T(Q)$ contains no Siegel or Cremer points. We claim that this is equivalent to the following:

(1) Q contains no Cremer point;
(2) if the boundary of a Siegel disk is decomposable, then Q is disjoint from it;
(3) if the boundary of a Siegel disk is indecomposable (it is not known if such Siegel disks exist), then Q intersects it in at most one point.

Indeed if (1) - (3) above are satisfied then, by an argument similar to the proof of Corollary 7.5.4, $T(Q)$ contains no Cremer or Siegel points. Now, suppose that $T(Q)$ contains no Cremer or Siegel points. Then by Corollary 7.5.4 $Q = \{q\}$ is a point. Hence (1) and (3) hold trivially. If B is the decomposable boundary of a Siegel disk and $q \in B$, then we may assume that B and Q are invariant. It is known [**Rog92a, Rog92b**] that there exists a monotone map $p : B \to \mathbb{S}^1$, an induced map $g : \mathbb{S}^1 \to \mathbb{S}^1$ which is an irrational rotation and for each $y \in \mathbb{S}^1$, at most one point of $p^{-1}(y)$ is accessible. Since Q is invariant, $Q = B$, a contradiction. Hence (2) holds as well.

Bibliography

[Aki99]   V. Akis, *On the plane fixed point problem*, Topology Proc. **24** (1999), 15–31. MR1802674 (2001k:54068)

[Bel67]   H. Bell, *On fixed point properties of plane continua*, Trans. A. M. S. **128** (1967), 539–548. MR0214036 (35:4888)

[Bel76]   _____, *A correction to my paper "Some topological extensions of plane geometry*, Rev. Colombiana Mat. **9** (1975), 125–153; Rev. Columbiana Mat. **10** (1976) 93. MR0467696 (57:7551a); MR0467697 (57:7551b)

[Bel78]   _____, *A fixed point theorem for plane homeomorphisms*, Fund. Math. **100** (1978), 119–128, See also: Bull. A. M. S. **82** (1976), 778-780. MR0500879 (58:18386); MR0410710 (53:14457)

[Bel79]   D. Bellamy, *A tree-like continuum without the fixed point property*, Houston J. of Math. **6** (1979), 1–13. MR575909 (81h:54039)

[Bin69]   R. H. Bing, *The elusive fixed point theorem*, Amer. Math. Monthly **76** (1969), 119–132. MR0236908 (38:5201)

[Bin81]   _____, *Commentary on Problem 107*, The Scottish Book, Birkhäuser, Boston, MA (1981), 190–192.

[Bor35]   K. Borsuk, *Einige Sätze über stetige Streckenbilder*, Fund. Math. **24** (1935), 51–58.

[BCLOS08] A. Blokh, D. Childers, G. Levin, L. Oversteegen, D. Schleicher, *Fatou–Shishikura inequality and wandering continua*, preprint arXiv:1001.0953, 58 pages (2010).

[BCO08]   A. Blokh, C. Curry, L. Oversteegen, *Locally connected models for Julia sets*, Advances in Mathematics, **226** (2011), 1621–1661. MR2737795

[BO09]   A. M. Blokh and L. G. Oversteegen, *A fixed point theorem for branched covering maps of the plane*, Fund. Math., **206** (2009), 77–111. MR2576262 (2011c:54043)

[BO06]   A. M. Blokh and L. G. Oversteegen, *Monotone images of Cremer Julia sets*, Houston Math. J., **36** (2010), 469–476. MR2661256 (2011e:37098)

[Bon04]   M. Bonino, *A Brouwer like theorem for orientation reversing homeomorphisms of the sphere*, Fund. Math. **182** (2004), 1–40. MR2100713 (2005m:37096)

[Bro12]   L. E. J. Brouwer, *Beweis des ebenen Translationessatzes*, Math. Ann. **72** (1912), 35–41.

[Bro77]   Morton Brown, *A short Proof of the Cartwright-Littlewood Theorem*, Proceedings AMS **65** (1977), 372. MR0461491 (57:1476)

[Bro84]   _____, *A new proof of Brouwer's lemma on translation arcs*, Houston J. of Math. **10** (1984), 35–41. MR736573 (85h:54080)

[Bro90]   _____, *Fixed point for orientation preserving homeomorphisms of the plane which interchange two points*, Pacific J. of Math., **143** (1990), 37–41. MR1047399 (91k:54072)

[CG93]   L. Carleson and T. W. Gamelin, *Complex dynamics*, Universitext: Tracts in Mathematics, Springer-Verlag, 1993. MR1230383 (94h:30033)

[CL51]   M. L. Cartwright and J. E. Littlewood, *Some fixed point theorems*, Annals of Math. (2) **54** (1951), 1–37. MR0042690 (13:148f)

[CL66]   E. F. Collingwood and A. J. Lohwater, *Theory of Cluster sets*, **56**, Cambridge Tracts in Math. and Math. Physics, Cambridge University Press, Cambridge, 1966. MR0231999 (38:325)

[Dav86]   Robert J. Daverman, *Decompositions of manifolds*, Academic Press, Inc., 1986. MR872468 (88a:57001)

[DH85a]   A. Douady and J. H. Hubbard, *Étude dynamique des polynômes complexes I, II* Publications Mathématiques d'Orsay **84-02** (1984), **85-04** (1985). MR0762431 (87f:58072a); MR0812271 (87f:58072b)

[DH85b]   A. Douady and J. H. Hubbard, *On the dynamics of polynomial-like mappings*, Ann. Sci. École Norm. Sup. (4) **18** (1985), no. 2, 287–343. MR816367 (87f:58083)
[Eng89]   R. Engelking, *General Topology*, Heldermann Verlag, 1989. MR1039321 (91c:54001)
[Fat87]   Albert Fathi, *An orbit closing proof of Brouwer's lemma on translation arcs*, L'enseignement Mathématique **33** (1987), 315–322. MR925994 (89d:55004)
[Fat20]   P. Fatou, *Sur les equations functionnelles*, Bull. Soc. Mat. France **48** (1920), 208–314.
[Fra92]   J. Franks, *A new proof of the Brouwer plane translation theorem*, Ergodic Theory and Dynamical Systems **12** (1992), 217–226. MR1176619 (93m:58059)
[Gui94]   L. Guillou, *Théorème de translation plane de Brouwer et généralisations du théorème de Poincaré-Birkhoff*, Topology **33** (1994), 331–351. MR1273787 (95h:55003)
[Hag71]   C. L. Hagopian, *A fixed-point theorem for plane continua*, Bull. Amer. Math. Soc. **77** (1971), 351–354. Addendum, ibid **78** (1972), 289. MR0273591 (42:8469); MR0288749 (44:5945)
[Hag96]   ———, *The fixed-point property for simply connected plane continua*, Trans. Amer. Math. Soc. **348** (1996), no. 11, 4525–4548. MR1344207 (97a:54047)
[Ham51]   O. H. Hamilton, *A fixed point theorem for pseudo-arcs and certain other metric continua* Trans. Amer. Math. Soc. **2** (1951), 18–24. MR0039993 (12:627f)
[Ili70]   S. D. Iliadis, *Location of continua on a plane and fixed points*, Vestnik Moskovskogo Univ. Matematika **25** (1970), no. 4, 66–70, Series I. MR0287522 (44:4726)
[Kiw04]   J. Kiwi, *Real laminations and the topological dynamics of complex polynomials*, Advances in Math. **184** (2004), no. 2, pp. 207–267. MR2054016 (2005b:37094)
[KP94]   Ravi S. Kulkarni and Ulrich Pinkall, *A canonical metric for Möbius structures and its applications*, Math. Z. **216(1)** (1994), 89–129. MR1273468 (95b:53017)
[KW91]   V. Klee and S. Wagon, *Old and new unsolved problems in plane geometry and number theory*, Dolciana Math. Expos. **11** (1991), Math. Assoc. Amer., Washington, DC. MR1133201 (92k:00014)
[Lel66]   A. Lelek, *On confluent mappings*, Colloq. Math. **15** (1966), 232–233. MR0208574 (34:8383)
[LR74]   A. Lelek and D. Read, *Compositions of confluent mappings and some other classes of functions*, Colloq. Math. **29** (1974), 101–112. MR0367900 (51:4142)
[LP96]   G. Levin and F. Przytycki, *External rays to periodic points,* Isr. J. Math. **94** (1996), 29–57. MR1394566 (97d:58164)
[Mil00]   J. Milnor, *Dynamics in one complex variable*, second ed., Vieweg, Wiesbaden, 2000. MR1721240 (2002i:37057)
[Min90]   P. Minc, *A fixed point theorem for weakly chainable plane continua*, Trans. Amer. Math. Soc. **317** (1990), 303–312. MR968887 (90d:54067)
[Min99]   ———, *A weakly chainable tree-like continuum without the fixed point property*, Trans. Amer. Math. Soc. **351** (1999), 1109–1121. MR1451610 (99e:54024)
[Nad92]   S. B. Nadler, Jr., *Continuum theory*, Marcel Dekker Inc., New York, 1992. MR1192552 (93m:54002)
[OT82]   L. Oversteegen, E. Tymchatyn, *Plane strips and the span of continua. I*, Houston J. Math. **8** (1982), no. 1, 129–142. MR666153 (84h:54030)
[OT07]   Lex G. Oversteegen and E.D. Tymchatyn, *Extending isotopies of planar continua*, Annals of Math., (2) **172** (2010), 2105–2133. MR2726106
[OV09]   Lex G. Oversteegen and Kirsten I. S. Valkenburg, *Characterizing isotopic continua in the sphere*, Proc. Amer. Math. Soc. **139** (2011), 1495–1510. MR2748444 (2012c:57041)
[Pom92]   Ch. Pommerenke, *Boundary behaviour of conformal maps*, **299**, Grundlehren der Math. Wissenschaften, Springer-Verlag, New York, 1992. MR1217706 (95b:30008)
[RY08]   P. Roesch, Y. Yin, *The boundary of bounded polynomial Fatou components*, Comptes Rendus Mathematique **346** (2008), 877-880. MR2441925 (2009k:37104)
[Rog92a] J. T. Rogers, Jr., *Is the boundary of a Siegel disk a Jordan curve?*, Bull. A. M. S., **27** (1992), 284–287. MR1160003 (93g:30009)
[Rog92b] J. T. Rogers, Jr., *Singularities in the boundaries of local Siegel disks*, Ergodic Theory and Dynamical Systems, **12** (1992), 803–821. MR1200345 (93m:58061)
[Rut35]   N. E. Rutt, *Prime ends and indecomposability*, Bull. A. M. S. **41** (1935), 265–273. MR1563071
[Shi87]   M. Shishikura, *On the quasiconformal surgery of rational functions*, Ann. Sci. Ecole Norm. Sup., **20** (1987) 1–29. MR892140 (88i:58099)

BIBLIOGRAPHY

[Sie05]   L. Siebenmann, *The Osgood-Schoenflies Theorem revisited*, Russian Math. Surveys, **60** (2005), 645–672. MR2190924 (2007e:57011)

[Sie68]   K. Sieklucki, *On a class of plane acyclic continua with the fixed point property*, Fund. Math. **63** (1968), 257–278. MR0240794 (39:2139)

[Slo91]   Z. Slodkowski, *Holomorphic motions and polynomial hulls*, Proc. AMS **111** (1991), 347–355. MR1037218 (91f:58078)

[Ste35]   Sternbach, Problem **107** (1935), in: *The Scottish Book: Mathematics from the Scottish Café*, Birkhäuser, Boston, 1981, 1935.

[ST86]   D. Sullivan and W. Thurston, *Extending holomorphic motions*, Acta Math. **157** (1986), 243–257. MR857674 (88i:30033)

[Thu09]   W. P. Thurston, *On the geometry and dynamics of iterated rational maps*, Complex dynamics, families and friends (Wellesley, MA, USA) (D. Schleicher, ed.), A. K. Peters, 2009, pp. 3–137. MR2508255 (2010m:37076)

[UY51]   H. D. Ursell and L. C. Young, *Remarks on the theory of prime ends*, Mem. of the Amer. Math. Soc. (1951), 29pp. MR0042110 (13:55a)

[Why42]   G. T. Whyburn, *Analytic topology* **28**, AMS Coll. Publications, Providence, RI, 1942. MR0007095 (4:86b)

[Why64]   ———, *Topological Analysis*, Princeton Math. Series. Princeton Univ. Press, Princeton, NJ, 1964. MR0165476 (29:2758)

Index

accessible point, 29
acyclic, 30
allowable partition, 28

$\mathfrak{B}$, 35
$\mathfrak{B}^\infty$, 35
boundary scrambling
  for dendrites, 50
  for planar continua, 51
branched covering map, 68
branchpoint of f, 75
bumping
  arc, 25
  simple closed curve, 25

$C(a,b)$, 38
Carathéodory Loop, 27
chain of crosscuts, 28
  equivalent, 28
channel, 29
  dense, 29, 57
completing a bumping arc, 25
$\mathbb{C}$, 1
$\mathbb{C}^\infty$, 1
conformal
  external ray, 53
continuum
  decomposable, 49
  indecomposable, 49
$\operatorname{conv}_{\mathcal{E}}(K)$, 35
$\operatorname{conv}_{\mathcal{H}}(B \cap K)$, 35
convex hull
  Euclidean, 35
  hyperbolic, 35
counterclockwise order
  on an arc in a simple closed curve, 13
critical point, 74
crosscut, 25
  shadow, 26
cutpoint, 51
C_X, 53, 85

$\mathbb{D}$, 38
defines variation near X, 50

degree, 13
$\deg_f(a)$, 75
$\operatorname{degree}(g)$, 13
$\operatorname{degree}(f_p)$, 16
∂ boundary operator, 2
$\mathbb{D}^\infty$, 27
dendrite, 6
domain
  attracting, 52
  Fatou, 52
  parabolic, 52
  Siegel, 52

embedding
  orientation preserving, 19
essential crossing, 29
$\mathcal{E}_t$, 29
exit continuum, 51
  for a general puzzle-piece, 53
external ray, 29
  end of, 29
  essential crossing, 29
  landing point, 29
  non smooth, 53
  smooth, 53

fixed point
  for positively oriented maps, 63
  non-rotational, 53
(f, X, η), 50

G, 38
$\mathfrak{g}$, 38
g, 35
Γ, 38
gap, 37, 68
general puzzle-piece, 53
geodesic
  hyperbolic, 38
geometric outchannel
  negative, 50
  positive, 50

hull
  hyperbolic, 37

INDEX

topological, 1
hyperbolic
   geodesic, 35
   halfplane, 35
hyperbolic chord, 38

id identity map, 13
$\text{Im}(\mathcal{E}_t)$, 29
impression, 29
index, 13
   fractional, 13
   I=V+1 for Carathéodory Loops, 28
   Index=Variation+1 Theorem, 20
   local, 52
$\text{ind}(f,x)$, 52
$\text{ind}(f,A)$, 19
$\text{ind}(f,g)$, 13
$\text{ind}(f,g|_{[a,b]})$, 13
$\text{ind}(f,S)$, 19

J_P, 52
Julia set, 52
junction, 14
Jørgensen Lemma, 39

K_P, 52
$\mathcal{KP}$, 37
$\mathcal{KP}_\delta$, 41
$\mathcal{KP}$-chord, 37
$\mathcal{KP}^\pm_\delta$, 43
$\mathcal{KPP}$, 37
$\mathcal{KPP}_\delta$, 41
Kulkarni-Pinkall
   Lemma, 36
   Partition, 37

lamination, 68
   degenerate, 68
   invariant, 68
landing point, 29
leaf, 68
link, 25
local degree, 75
local index, 52
Lollipop Lemma, 23

map, 13
   confluent, 16
   light, 16
   monotone, 16
   negatively oriented, 16
   on circle of prime ends, 32
   oriented, 16
   perfect, 16
   positively oriented, 16
maximal ball, 35
Maximum Modulus Theorem, 31
monotone-light decomposition of a map, 16

narrow strip, 59

natural retraction of dendrites, 64
non-separating, xi

order
   on subarc of simple closed curve, 13
orientation preserving
   embedding, 19
outchannel, 50
   geometric, 50
   uniqueness, 59

periodic point
   Cremer, 53
   parabolic, 52
   repelling, 52
   Siegel, 52
   weakly repelling, 51
point of period two
   for oriented maps, 64
positively oriented arc, 77
$\text{Pr}(\mathcal{E}_t)$, 29
prime end, 28
   channel, 29
   impression, 29
   principal continuum, 29
principal continuum, 29

$\mathbb{R}$, 1
repels outside X at p, 76
R_t, 29

shadow, 26
$\text{Sh}(A)$, 26
smallest ball, 35
standing hypothesis, 50

topological
   Julia set, 68
   polynomial, 68
topological hull, 1
topologically
   attracting, 75
   repelling, 75
$T(X)$, 1
$T(X)_\delta$, 43

U^∞, 1
unlinked, 68

$\text{val}_Y(x)$, 51
valence, 51
$\text{var}(f,A)$, 27
variation
   for crosscuts, 25
   of a simple closed curve, 15
   on an arc, 14
   on finite union of arcs, 15
$\text{var}(f,A,S)$, 14
$\text{var}(f,S)$, 15

weakly repelling, 51, 65
wedge (at an exit continuum), 53
win$(g, \mathbb{S}^1, w)$, 13

Editorial Information

To be published in the *Memoirs*, a paper must be correct, new, nontrivial, and significant. Further, it must be well written and of interest to a substantial number of mathematicians. Piecemeal results, such as an inconclusive step toward an unproved major theorem or a minor variation on a known result, are in general not acceptable for publication.

Papers appearing in *Memoirs* are generally at least 80 and not more than 200 published pages in length. Papers less than 80 or more than 200 published pages require the approval of the Managing Editor of the Transactions/Memoirs Editorial Board. Published pages are the same size as those generated in the style files provided for $\mathcal{AMS}$-LATEX or $\mathcal{AMS}$-TEX.

Information on the backlog for this journal can be found on the AMS website starting from http://www.ams.org/memo.

A Consent to Publish is required before we can begin processing your paper. After a paper is accepted for publication, the Providence office will send a Consent to Publish and Copyright Agreement to all authors of the paper. By submitting a paper to the *Memoirs*, authors certify that the results have not been submitted to nor are they under consideration for publication by another journal, conference proceedings, or similar publication.

Information for Authors

Memoirs is an author-prepared publication. Once formatted for print and on-line publication, articles will be published as is with the addition of AMS-prepared frontmatter and backmatter. Articles are not copyedited; however, confirmation copy will be sent to the authors.

Initial submission. The AMS uses Centralized Manuscript Processing for initial submissions. Authors should submit a PDF file using the Initial Manuscript Submission form found at www.ams.org/submission/memo, or send one copy of the manuscript to the following address: Centralized Manuscript Processing, MEMOIRS OF THE AMS, 201 Charles Street, Providence, RI 02904-2294 USA. If a paper copy is being forwarded to the AMS, indicate that it is for *Memoirs* and include the name of the corresponding author, contact information such as email address or mailing address, and the name of an appropriate Editor to review the paper (see the list of Editors below).

The paper must contain a *descriptive title* and an *abstract* that summarizes the article in language suitable for workers in the general field (algebra, analysis, etc.). The *descriptive title* should be short, but informative; useless or vague phrases such as "some remarks about" or "concerning" should be avoided. The *abstract* should be at least one complete sentence, and at most 300 words. Included with the footnotes to the paper should be the 2010 *Mathematics Subject Classification* representing the primary and secondary subjects of the article. The classifications are accessible from www.ams.org/msc/. The Mathematics Subject Classification footnote may be followed by a list of *key words and phrases* describing the subject matter of the article and taken from it. Journal abbreviations used in bibliographies are listed in the latest *Mathematical Reviews* annual index. The series abbreviations are also accessible from www.ams.org/msnhtml/serials.pdf. To help in preparing and verifying references, the AMS offers MR Lookup, a Reference Tool for Linking, at www.ams.org/mrlookup/.

Electronically prepared manuscripts. The AMS encourages electronically prepared manuscripts, with a strong preference for $\mathcal{AMS}$-LATEX. To this end, the Society has prepared $\mathcal{AMS}$-LATEX author packages for each AMS publication. Author packages include instructions for preparing electronic manuscripts, samples, and a style file that generates the particular design specifications of that publication series. Though $\mathcal{AMS}$-LATEX is the highly preferred format of TEX, author packages are also available in $\mathcal{AMS}$-TEX.

Authors may retrieve an author package for *Memoirs of the AMS* from www.ams.org/journals/memo/memoauthorpac.html or via FTP to ftp.ams.org (login as anonymous, enter your complete email address as password, and type cd pub/author-info). The

AMS Author Handbook and the *Instruction Manual* are available in PDF format from the author package link. The author package can also be obtained free of charge by sending email to `tech-support@ams.org` or from the Publication Division, American Mathematical Society, 201 Charles St., Providence, RI 02904-2294, USA. When requesting an author package, please specify $\mathcal{AMS}$-LaTeX or $\mathcal{AMS}$-TeX and the publication in which your paper will appear. Please be sure to include your complete mailing address.

After acceptance. The source files for the final version of the electronic manuscript should be sent to the Providence office immediately after the paper has been accepted for publication. The author should also submit a PDF of the final version of the paper to the editor, who will forward a copy to the Providence office.

Accepted electronically prepared files can be submitted via the web at `www.ams.org/submit-book-journal/`, sent via FTP, or sent on CD to the Electronic Prepress Department, American Mathematical Society, 201 Charles Street, Providence, RI 02904-2294 USA. TeX source files and graphic files can be transferred over the Internet by FTP to the Internet node `ftp.ams.org` (130.44.1.100). When sending a manuscript electronically via CD, please be sure to include a message indicating that the paper is for the *Memoirs*.

Electronic graphics. Comprehensive instructions on preparing graphics are available at `www.ams.org/authors/journals.html`. A few of the major requirements are given here.

Submit files for graphics as EPS (Encapsulated PostScript) files. This includes graphics originated via a graphics application as well as scanned photographs or other computer-generated images. If this is not possible, TIFF files are acceptable as long as they can be opened in Adobe Photoshop or Illustrator.

Authors using graphics packages for the creation of electronic art should also avoid the use of any lines thinner than 0.5 points in width. Many graphics packages allow the user to specify a "hairline" for a very thin line. Hairlines often look acceptable when proofed on a typical laser printer. However, when produced on a high-resolution laser imagesetter, hairlines become nearly invisible and will be lost entirely in the final printing process.

Screens should be set to values between 15% and 85%. Screens which fall outside of this range are too light or too dark to print correctly. Variations of screens within a graphic should be no less than 10%.

Inquiries. Any inquiries concerning a paper that has been accepted for publication should be sent to `memo-query@ams.org` or directly to the Electronic Prepress Department, American Mathematical Society, 201 Charles St., Providence, RI 02904-2294 USA.

Editors

This journal is designed particularly for long research papers, normally at least 80 pages in length, and groups of cognate papers in pure and applied mathematics. Papers intended for publication in the *Memoirs* should be addressed to one of the following editors. The AMS uses Centralized Manuscript Processing for initial submissions to AMS journals. Authors should follow instructions listed on the Initial Submission page found at www.ams.org/memo/memosubmit.html.

Algebra, to ALEXANDER KLESHCHEV, Department of Mathematics, University of Oregon, Eugene, OR 97403-1222; e-mail: klesh@uoregon.edu

Algebraic geometry, to DAN ABRAMOVICH, Department of Mathematics, Brown University, Box 1917, Providence, RI 02912; e-mail: amsedit@math.brown.edu

Algebraic topology, to SOREN GALATIUS, Department of Mathematics, Stanford University, Stanford, CA 94305 USA; e-mail: transactions@lists.stanford.edu

Arithmetic geometry, to TED CHINBURG, Department of Mathematics, University of Pennsylvania, Philadelphia, PA 19104-6395; e-mail: math-tams@math.upenn.edu

Automorphic forms, representation theory and combinatorics, to DANIEL BUMP, Department of Mathematics, Stanford University, Building 380, Sloan Hall, Stanford, California 94305; e-mail: bump@math.stanford.edu

Combinatorics, to JOHN R. STEMBRIDGE, Department of Mathematics, University of Michigan, Ann Arbor, Michigan 48109-1109; e-mail: JRS@umich.edu

Commutative and homological algebra, to LUCHEZAR L. AVRAMOV, Department of Mathematics, University of Nebraska, Lincoln, NE 68588-0130; e-mail: avramov@math.unl.edu

Differential geometry and global analysis, to CHRIS WOODWARD, Department of Mathematics, Rutgers University, 110 Frelinghuysen Road, Piscataway, NJ 08854; e-mail: ctw@math.rutgers.edu

Dynamical systems and ergodic theory and complex analysis, to YUNPING JIANG, Department of Mathematics, CUNY Queens College and Graduate Center, 65-30 Kissena Blvd., Flushing, NY 11367; e-mail: Yunping.Jiang@qc.cuny.edu

Functional analysis and operator algebras, to NATHANIEL BROWN, Department of Mathematics, 320 McAllister Building, Penn State University, University Park, PA 16802; e-mail: nbrown@math.psu.edu

Geometric analysis, to WILLIAM P. MINICOZZI II, Department of Mathematics, Johns Hopkins University, 3400 N. Charles St., Baltimore, MD 21218; e-mail: trans@math.jhu.edu

Geometric topology, to MARK FEIGHN, Math Department, Rutgers University, Newark, NJ 07102; e-mail: feighn@andromeda.rutgers.edu

Harmonic analysis, complex analysis, to MALABIKA PRAMANIK, Department of Mathematics, 1984 Mathematics Road, University of British Columbia, Vancouver, BC, Canada V6T 1Z2; e-mail: malabika@math.ubc.ca

Harmonic analysis, representation theory, and Lie theory, to E. P. VAN DEN BAN, Department of Mathematics, Utrecht University, P.O. Box 80 010, 3508 TA Utrecht, The Netherlands; e-mail: E.P.vandenBan@uu.nl

Logic, to ANTONIO MONTALBAN, Department of Mathematics, The University of California, Berkeley, Evans Hall #3840, Berkeley, California, CA 94720; e-mail: antonio@math.berkeley.edu

Number theory, to SHANKAR SEN, Department of Mathematics, 505 Malott Hall, Cornell University, Ithaca, NY 14853; e-mail: ss70@cornell.edu

Partial differential equations, to GUSTAVO PONCE, Department of Mathematics, South Hall, Room 6607, University of California, Santa Barbara, CA 93106; e-mail: ponce@math.ucsb.edu

Partial differential equations and functional analysis, to ALEXANDER KISELEV, Department of Mathematics, University of Wisconsin-Madison, 480 Lincoln Dr., Madison, WI 53706; e-mail: kisilev@math.wisc.edu

Probability and statistics, to PATRICK FITZSIMMONS, Department of Mathematics, University of California, San Diego, 9500 Gilman Drive, La Jolla, CA 92093-0112; e-mail: pfitzsim@math.ucsd.edu

Real analysis and partial differential equations, to WILHELM SCHLAG, Department of Mathematics, The University of Chicago, 5734 South University Avenue, Chicago, IL 60615; e-mail: schlag@math.uchicago.edu

All other communications to the editors, should be addressed to the Managing Editor, ALEJANDRO ADEM, Department of Mathematics, The University of British Columbia, Room 121, 1984 Mathematics Road, Vancouver, B.C., Canada V6T 1Z2; e-mail: adem@math.ubc.ca

Selected Published Titles in This Series

1051 **Ariel Barton,** Elliptic Partial Differential Equations with Almost-Real Coefficients, 2013

1050 **Thomas Lam, Luc Lapointe, Jennifer Morse, and Mark Shimozono,** The Poset of k-Shapes and Branching Rules for k-Schur Functions, 2013

1049 **David I. Stewart,** The Reductive Subgroups of F_4, 2013

1048 **Andrzej Nagórko,** Characterization and Topological Rigidity of Nöbeling Manifolds, 2013

1047 **Joachim Krieger and Jacob Sterbenz,** Global Regularity for the Yang-Mills Equations on High Dimensional Minkowski Space, 2013

1046 **Keith A. Kearnes and Emil W. Kiss,** The Shape of Congruence Lattices, 2013

1045 **David Cox, Andrew R. Kustin, Claudia Polini, and Bernd Ulrich,** A Study of Singularities on Rational Curves Via Syzygies, 2013

1044 **Steven N. Evans, David Steinsaltz, and Kenneth W. Wachter,** A Mutation-Selection Model with Recombination for General Genotypes, 2013

1043 **A. V. Sobolev,** Pseudo-Differential Operators with Discontinuous Symbols: Widom's Conjecture, 2013

1042 **Paul Mezo,** Character Identities in the Twisted Endoscopy of Real Reductive Groups, 2013

1041 **Verena Bögelein, Frank Duzaar, and Giuseppe Mingione,** The Regularity of General Parabolic Systems with Degenerate Diffusion, 2013

1040 **Weinan E and Jianfeng Lu,** The Kohn-Sham Equation for Deformed Crystals, 2013

1039 **Paolo Albano and Antonio Bove,** Wave Front Set of Solutions to Sums of Squares of Vector Fields, 2013

1038 **Dominique Lecomte,** Potential Wadge Classes, 2013

1037 **Jung-Chao Ban, Wen-Guei Hu, Song-Sun Lin, and Yin-Heng Lin,** Zeta Functions for Two-Dimensional Shifts of Finite Type, 2013

1036 **Matthias Lesch, Henri Moscovici, and Markus J. Pflaum,** Connes-Chern Character for Manifolds with Boundary and Eta Cochains, 2012

1035 **Igor Burban and Bernd Kreussler,** Vector Bundles on Degenerations of Elliptic Curves and Yang-Baxter Equations, 2012

1034 **Alexander Kleshchev and Vladimir Shchigolev,** Modular Branching Rules for Projective Representations of Symmetric Groups and Lowering Operators for the Supergroup $Q(n)$, 2012

1033 **Daniel Allcock,** The Reflective Lorentzian Lattices of Rank 3, 2012

1032 **John C. Baez, Aristide Baratin, Laurent Freidel, and Derek K. Wise,** Infinite-Dimensional Representations of 2-Groups, 2012

1031 **Idrisse Khemar,** Elliptic Integrable Systems: A Comprehensive Geometric Interpretation, 2012

1030 **Ernst Heintze and Christian Groß,** Finite Order Automorphisms and Real Forms of Affine Kac-Moody Algebras in the Smooth and Algebraic Category, 2012

1029 **Mikhail Khovanov, Aaron D. Lauda, Marco Mackaay, and Marko Stošić,** Extended Graphical Calculus for Categorified Quantum sl(2), 2012

1028 **Yorck Sommerhäuser and Yongchang Zhu,** Hopf Algebras and Congruence Subgroups, 2012

1027 **Olivier Druet, Frédéric Robert, and Juncheng Wei,** The Lin-Ni's Problem for Mean Convex Domains, 2012

1026 **Mark Behrens,** The Goodwillie Tower and the EHP Sequence, 2012

1025 **Joel Smoller and Blake Temple,** General Relativistic Self-Similar Waves that Induce an Anomalous Acceleration into the Standard Model of Cosmology, 2012

For a complete list of titles in this series, visit the
AMS Bookstore at **www.ams.org/bookstore/memoseries/**.